Beginner's Guide to SPE

Solid-Phase Extraction

Joseph C. Arsenault

THE SCIENCE OF WHAT'S POSSIBLE.™

Copyright © 2012 Waters Corporation

All rights reserved. No part of this book may be reproduced or transmitted in any form by any means, electronic or mechanical, including photocopying, recording, or by any other information storage and retrieval system, without permission from the Publisher.

Waters Corporation

34 Maple Street

Milford, MA 01757

Library of Congress Control Number: 2012944477

Printed in the USA

©2012 Waters Corporation. Waters, Sep-Pak, Oasis, UPLC, Nova-Pak, PoraPak, ACQUITY UPLC, ACQUITY, Atlantis, Xevo, and Alliance are registered trademarks of Waters Corporation. XPoSure, DisQuE, AccQ•Tag, Quattro micro GC, and SunFire are trademarks of Waters Corporation. Luer is a registered trademark of Becton Dickinson. Teflon is a registered trademark of E. I. du Pont de Nemours and Company. Kool-Aid is a registered trademark of Kraft Foods Global Brands, LLC. Rxni is a registered trademark of Rxi Pharmaceuticals Corporation.

October 2012 715003405 VW-FP

INTRODUCTION

Welcome!

We are pleased that you are interested in learning more about solid-phase extraction [SPE] technology. We have developed this text to show you how to put SPE to work in your laboratory to solve sample preparation problems. With advances in today's analytical equipment, especially in the areas of speed and sensitivity, more demands are being placed on improving how we prepare samples for analysis.

SPE is somewhat like the proverbial "Genie in the Bottle." Once you understand how to use its power, you will be amazed at the types of analytical problems it can help you solve. As with the Genie, SPE is a tool, and its power is determined by the creativity of the "Master," you the analytical scientist.

In this book, we have endeavored to provide all of the SPE fundamentals and success techniques derived from scientists from all over the world who have counted on this technology in the past thirty years. Today, scientists are finding SPE more useful than ever in solving difficult sample preparation and analytical problems. We hope this book will enable you to understand and master the capabilities of SPE, so that you too can put the power of this technology to use in your laboratory.

We wish you great success.

DEDICATIONS

I would like to make a few, heartfelt dedications.

First, to Dr. Patrick McDonald, here at Waters Corporation, who pioneered the field of SPE over 34 years ago, with the development of Sep-Pak® cartridges, and then the creation of the Oasis® family of SPE sorbents. As a superb scientist, he has always used his knowledge and wisdom to overcome problems facing analytical scientists around the world by helping develop innovative chromatographic products, such as SPE technology.

My second dedication is to Dr. Uwe Neue, one of the true gurus of chromatography. Dr. Neue passed away in 2010 leaving an incredible legacy of understanding and innovation in the world of liquid chromatography.

My success was in great part due to both these individuals who shared their knowledge and helped me, and the rest of the world, learn to harness the power of chromatographic science.

Finally, I am so pleased to dedicate this publication to my wonderful family, who, with me, grew up with liquid chromatography. My wife Ann Marie, my daughter Christine, and my son Joseph, inspired me to learn the science, and help colleagues and customers better understand how to improve our lives today, and create a better world for future generations.

ACKNOWLEDGEMENTS

I would like to thank Dr. Diane Diehl and Dr. Mark Baynham whose vision and support were instrumental in the development of this book.

To Kathy Coffey, Jessalynn Wheaton, John Martin, Dr. Nebila Idris, Damian Morrison, Erin Chambers, Dr. Michael S. Young, Dr. James Teuscher, Dr. Ben Yong, Xin Zhang, Nicole Cotta, and Gary Mantha for their assistance and comments.

And finally, my great appreciation to the Waters Chemistry Marketing Communications Team, especially to Natalie Crosier, Vicki Walton, and Ian Hanslope, who worked so diligently in the production of this book.

TABLE OF CONTENTS

INTRODUCTION 3

DEDICATIONS AND ACKNOWLEDGEMENTS 4

BENEFITS OF SOLID-PHASE EXTRACTION IN SAMPLE PREPARATION 17

Definiton of Solid-Phase Extraction 18

Four Major Benefits of SPE 19

    1. Simplification of Complex Sample Matrix along with Compound Purification 19

    2. Reduce Ion Suppression or Enhancement in MS Applications 22

    3. Capability to Fractionate Sample Matrix to Analyze Compounds by Class 23

    4. Trace Concentration [Enrichment] of Very Low Level Compounds 25

SPE IS LC 27

Separation Based on Polarity 29

    Designing a Polarity—Based Method 30

    Mobile Phase 30

    Stationary Phase 31

    Analyte Characteristics 31

    Normal-Phase SPE 32

    Reversed-Phase SPE 33

    Importance of Conditioning and Equilibration Steps 35

    The Effect of De-Wetting Phenomena/Drying Out Effect on SPE 39

    Retention Mechanism Map for Reversed-Phase SPE 41

    The Power of pH 43

Separations Based on Charge: Ion-Exchange Chromatography [IEC] 44

    Ionization States 44

    Choosing an Ion Exchange Sorbent for SPE 48

    Strong Cation-Exchange and Reversed-Phase Sorbent 48

    Strong Anion-Exchange and Reversed-Phase Sorbent 49

    Weak Cation-Exchange and Reversed-Phase Sorbent 50

[Table of Contents]

    Weak Anion-Exchange and Reversed-Phase Sorbent 51

    A Powerful Strategy to Simplify SPE Method Development Using Four Ion-Exchange Sorbents 51

Specialty Techniques 55

    Sep-Pak DNPH-Silica Cartridges for Analyzing Formaldehyde Other Aldehydes and Ketones in Air 55

    Ozone Scrubber Cartridges 57

    Waters XPoSure Aldehyde Sampler Cartridges for Monitoring Aldehydes in Indoor
    Air with Personal Portable Pumps 58

    PoraPak Rdx Sep-Pak Extraction Cartridge for the Analysis of
    Explosives in Surface and Ground Waters 59

    Sep-Pak Dry SPE Cartridge 60

KEY TERMS AND CALCULATIONS 61

Determining the Hold-Up Volume 63

Flow Rate and Linear Velocity 64

    Understanding Linear Velocity 64

    Importance of Flow Rate Control 65

    Scaling and Linear Velocity 66

Calculating Recovery 69

    Adjusting for Errors from the SPE Device or Protocol 70

Matrix Effects Calculations 72

IN THE LAB 77

Solid-Phase Extraction Device Designs 78

Syringe Style 79

    Syringe Barrel Design 79

    Vac RC design 80

96-Well Plate Designs 81

    µElution 96-Well Plate Design 82

Enclosed Design Cartridges 83

    Classic Design 83

    Plus Design 84

    Light Design 84

On-Line Processing..85

Dispersive SPE..87

Basic Set Up-Driving Forces Used to Generate Liquid Flow 90
    Gravity..90
    Positive Displacement or Pump Pressure 92
    Vacuum..95
        Applying Samples and Solvents to Cartridges...97
        Using Vacuum with Enclosed Cartridges 99
        Using Vacuum with 96-Well Plates..100

Steps in an SPE Method..101
    Pretreatment of the Original Sample 101
    Preparing the Cartridge 102
    Condition the SPE Cartridge..102
    Equilibrate the SPE Cartridge 104
    Load the Sample 106
    Wash Away Interference[s] 108
    Elute the Compound[s] of Interest 109
    Dry Down and Reconstitute in Mobile Phase..110

METHOD DEVELOPMENT 113

Sample/Analyte Attributes..115

Sample Matrix Pretreatment 115

SPE Strategies to Solve Different Problems 116
    Strategy 1: Pass Through 116
        1. Sample Volume Greater Than Hold-Up Volume 117
        2. Sample Volume Less Than Hold-Up Volume 119
    Strategy 2: Capture the Compound[s] of Interest...122
    Strategy 3: Capture and Fractionate the Compounds of Interest 125
    Strategy 4: Capture and Trace Enrich [Concentrate] the Compound[s] of Interest 131
        Trace Enriching Method..131
        Calculating the Results 137
    Chromatographic Mode Selection 138
    Choosing the Right Device Size 139

[Table of Contents]

Mass Balance in Method Development and Troubleshooting 139
Choosing Your Strategy 144
    Strategy 1: Pass Through 144
    Strategy 2: Capture 147
    Strategy 3: Capture and Fractionate 150
    Strategy 4: Capture and Trace Enrichment 151

Load Capacity/Breakthrough 154
    Importance of Chromatographic Conditions 155
    Breakthrough Study Experiment 156
    Scaling of the Method Volumes and Sorbent Mass 162

TROUBLESHOOTING 165

General Problem Sources in Sample Preparation 168
    Temperature Control 168
    Flow Rate Control 168
    pH Control 168
    Pretreatment and Extraction 171
    Sample Matrix 172
    Vial Condition 172
    Sample Solvent 172

APPENDIX: GLOSSARY OF SPE & LC TERMS 173

APPENDIX: OASIS SORBENT TECHNOLOGY FOR SPE 191

Introduction 192
A Wide Selection of Oasis Chemistries 193
High and Consistent Recoveries 194
Effect of Drying on Recovery—Oasis HLB Versus C_{18} Sorbents 195
High Capacity Using Less Sorbent 196
Exceptional Batch-to-Batch Reproducibility 196
Sorbent Amount and Solvent Selection for the Generic SPE Method 197
Simplified Method Development Protocol 198

APPENDIX: APPLICATIONS 201

Determination of PAH in Seafood 202
Melamine and Cyanuric Acid in Infant Formula 203
Multi-Residue Determination of Veterinary Drugs in Milk 204
Amyloid β Peptide in Cerebrospinal Fluid 204
Removal of Polyethylene Glycol 400 (PEG 400) from Plasma 205
Plasma and Serum Protein Digests 205
Endocrine-Disrupting Compounds in River Water 206
Organochlorine Pesticides and PCBs in Soil 206

APPENDIX: ADDITIONAL REFERENCE MATERIALS 207

[List of Figures]

LIST OF FIGURES

Figure 1: Examples of an SPE Method 19
Figure 2: Example of a Complex Sample 20
Figure 3: Comparison of Sample Matrix Complexities 20
Figure 4: Improved Quantitation with Better Sample Preparation 21
Figure 5: Significant Improvement in Baseline Using SPE Technology 21
Figure 6: Example of Ion Suppression Due to Sample Matrix 22
Figure 7: Reduced Ion Suppression with Proper SPE 23
Figure 8: Sample Preparation by SPE 24
Figure 9: Example of Trace Concentration 25
Figure 10: The Power of SPE 26
Figure 11: Polarity Spectrum 29
Figure 12: Proper Combination of Mobile and Stationary Phases Affects Separation Based on Polarity 30
Figure 13: Mobile Phase Chromatographic Polarity Spectrum 30
Figure 14: Stationary-Phase Particle Chromatographic Polarity Spectrum 31
Figure 15: Compound/Analyte Chromatographic Polarity Spectrum 31
Figure 16: Normal-Phase Chromatography 32
Figure 17: Reversed-Phase Chromatography, SPE and LC 33
Figure 18: Reversed-Phase SPE 34
Figure 19: Polar Surface with Polar Liquid 35
Figure 20: Non-Polar Surface and Polar Liquid 36
Figure 21: Reversed-Phase Sorbent with Wetted Pore 37
Figure 22: Pore De-Wetting 38
Figure 23: Loss of Analyte Retention Due to De-Wetted Pores 38
Figure 24: Conditioning and Equilibration Steps 39
Figure 25: Poor % Recovery Due to De-Wetting 40
Figure 26: Retention Map for Oasis HLB Pure Reversed Phase 42
Figure 27: Ion-Exchange Chromatography 44

List of Figures

Figure 28: Charged State of Analytes vs. pH 45

Figure 29: Oasis Family of Ion-Exchange Sorbents [RP = Reversed Phase] 45

Figure 30: Oasis MCX Retention Map for a Weak Base 48

Figure 31: Oasis MAX Retention Map for Weak Acid 49

Figure 32: Oasis WCX Retention Map for Strong Base 50

Figure 33: Oasis WAX Retention Map for Strong Acid 51

Figure 34: Oasis 2x4 Method Development Strategy 52

Figure 35: Charged States of Analytes Relative to pH 52

Figure 36: Capture of a Weak Base on Oasis MCX 53

Figure 37: Capture of a Strong Acid on Oasis WAX 53

Figure 38: Capture of a Weak Acid on Oasis MAX 54

Figure 39: Capture of a Strong Base on Oasis WCX 54

Figure 40: Capture of a Zwitterion on Oasis MCX 55

Figure 41: Sep-Pak DNPH Cartridges—Short and Long Body 56

Figure 42: HPLC Separation of DNPH Derivatives of Common Aldehydes and Ketones [DNPH] 56

Figure 43: Low-Level: Aldehyde Profile from Laboratory Air 57

Figure 44: High-Level: Aldehyde Profile from Diluted Auto Exhaust Emissions 57

Figure 45: Ozone Scrubber Cartridge 58

Figure 46: Flow Schematic for Air Sampling System 58

Figure 47: Waters XPoSure Aldehyde Sampler 58

Figure 48: Low-Level Example: Aldehyde Profile from Laboratory Air Using XPoSure Aldehyde Samplers 59

Figure 49: PoraPak Rdx Cartridges 59

Figure 50: Isocratic Separation of Method 8330 Analytes [PoraPak Rdx] 60

Figure 51: Sep-Pak Dry Cartridge 60

Figure 52: SPE Device Hold-Up Volume 62

Figure 53: Familiar Example of Interstitial Volume 63

Figure 54: Load, Wash, and Elution Steps Require Flow Rate Control 65

Figure 55: Importance of Flow Rate Control—Load Steps 66

Figure 56: Scaling from Smaller to Larger SPE Device 67

[List of Figures]

Figure 57: Scaling from Large to Smaller Device 68
Figure 58: Determining % Recovery 69
Figure 59: Proper Determination of % Recovery 70
Figure 60: Calculating % Recovery 71
Figure 61: Proper Determination for Matrix Effects 73
Figure 62: First Scenario: No Matrix Effects 74
Figure 63: Second Scenario: Significant Ion Suppression 75
Figure 64: Third Scenario: Ion Enhancement 76
Figure 65: SPE Device Configurations 78
Figure 66: Diagram of Syringe Barrel Design 80
Figure 67: Expanded Reservoir Barrel Design, Vac RC 80
Figure 68: 96-Well Plates 81
Figure 69: 96-Well Plate Bed Configurations for Oasis Sorbents 81
Figure 70: 96-Well Plate Bed Configuations for Silica-Based Sorbents 82
Figure 71: Waters Oasis µElution Plate 82
Figure 72: Patented µElution Plate Tip Design 83
Figure 73: Classic Cartridge 83
Figure 74: Plus Cartridges 84
Figure 75: Light Cartridge 84
Figure 76: On-line SPE Devices 85
Figure 77: Flow Diagrams of an On-line SPE System During the Load and Elution Steps 86
Figure 78: A Complex On-line SPE System with 2 SPE Devices 87
Figure 79: DisQuE Dispersive Sample Preparation for QuEChERS 87
Figure 80: DisQuE Tube 1 Protocol [QuEChERS] 88
Figure 81: DisQuE Tube 2 Protocol [QuEChERS] 89
Figure 82: Using Gravity as a Driving Force 90
Figure 83: Using Gravity as a Driving Force 91
Figure 84: Using Centrifugal Force 91
Figure 85: Using a Syringe/Plunger to Create Positive Displacement as a Driving Force 92

[List of Figures]

Figure 86: Proper Flow Rate Control—Drops 92

Figure 87: Importance of Flow Rate Control 93

Figure 88: Adapting a Syringe for Positive Displacement 93

Figure 89: Adapting an Enclosed Cartridge for a Peristaltic Pump 94

Figure 90: Positive Pressure Processor for 96-Well Plates 94

Figure 91: Glass Vacuum Manifold 95

Figure 92: Set up of Glass Vacuum Manifold 96

Figure 93: Applying Liquids to Syringe Design 97

Figure 94: Adapting Resevoirs 98

Figure 95: Applying a Large Volume Sample 98

Figure 96: Enclosed SPE Devices 99

Figure 97: Applying Liquids to Enclosed SPE Devices 99

Figure 98: 96-Well Plate on Manifold 100

Figure 99: Sorbent Pores must be Properly Wetted 103

Figure 100: Conditioning the Pores of a Reversed-Phase Sorbent with Organic Solvent 104

Figure 101: Equilibrating the Pores with Water or Sample Solvent 105

Figure 102: Proper Wetting of Pores is Critical for Capture During Loading Steps 106

Figure 103: Load the Sample 107

Figure 104: Wash Away Interferences 108

Figure 105: Elute the Compounds of Interest 109

Figure 106: Dry Down and Reconstitute in LC Mobile Phase 110

Figure 107: Why Dry Down and Reconstitution Steps are Required 111

Figure 108: Strategy 1: Pass Through 117

Figure 109: Pass-Through Strategy when the Sample Volume is Greater than the Hold-up Volume 118

Figure 110: Pass-Through Strategy when the Sample Volume is Greater than the Hold-up Volume 119

Figure 111: Pass-Through Strategy when the Sample Volume is Less Than the Hold-up Volume 120

Figure 112: Pass-Through Strategy when the Sample Volume is Less Than the Hold-up Volume 121

Figure 113: Capture Strategy—Load Step 122

Figure 114: Strategy 2: Capture Strategy—Wash Step 123

[List of Figures]

Figure 115: Capture Strategy—Elution Step, Second Case 124

Figure 116: Capture and Fractionate Strategy 125

Figure 117: Capture and Fractionate Strategy—Load Step 126

Figure 118: Capture and Fractionate Strategy—Elute 1 127

Figure 119: Capture and Fractionate Strategy—Elute 2 128

Figure 120: Capture and Fractionate Strategy—Elute 3 129

Figure 121: Capture and Fractionate Strategy—Elute 4 130

Figure 122: Using SPE for Trace Enrichment—Strategy 4 132

Figure 123: Trace Enrichment Strategy—Additional Sample Volume Loaded 133

Figure 124: Trace Enrichment Strategy—Elution Step 134

Figure 125: Trace Enrichment Strategy—Reversed Flow Elution Step—Syringe Design 135

Figure 126: Trace Enrichment Strategy—Reversed Flow Elution Step—Plus Design 136

Figure 127: Mass Balance—Impact of k Value 140

Figure 128: Mass Balance—Changing the k Value 141

Figure 129: Mass Balance—Changing Sample Volume 142

Figure 130: Benefits of Strong Elution Solvents 143

Figure 131: Mass Balance—Pass-Through Strategy—Complete Elution Step 145

Figure 132: Mass Balance—Pass-Through Strategy—Incomplete Elution Steps 146

Figure 133: Mass Balance—Capture 147

Figure 134: Mass Balance—Capture—Complete Elution Step 149

Figure 135: Mass Balance—Capture and Fractionate 150

Figure 136: Mass Balance–Trace Enrichment 152

Figure 137: Reversed Flow Direction for Elution Step 154

Figure 138: Breakthrough Study—Loading 156

Figure 139: Breakthough Study: Recovery vs. Load Volume 157

Figure 140: % Recovery vs. Load Volume 160

Figure 141: Loading Step 161

Figure 142: Retention Map for Pure Reversed-Phase Sorbent 168

Figure 143: Effect of pH on Silica-Based SPE Sorbents 169

List of Figures

Figure 144: Retention Map for Silica C_{18} Sorbent 169

Figure 145: Sorbent Charge States 170

Figure 146: Poor Recovery of Bases Using Silica-Based Sorbents When pH is Not Controlled 171

Figure 147: Methods for Calculating Plate Number [N] 177

Figure 148: Elution Strength for Reversed Phase 179

Figure 149: High Pressure Gradient System 180

Figure 150: Low Pressure Gradient System 181

Figure 151: Hold-Up Volume—SPE Cartridge 182

Figure 152: Methods for Calculating Plate Number [N] 186

Figure 153: Unique Water-Wettable Oasis HLB Copolymer 193

Figure 154: Oasis Sorbents 194

Figure 155: Oasis HLB: No Impact of Sorbent Drying Out 195

Figure 156: Higher Retention Means Greater Capacity, No Breakthrough 196

Figure 157: Batch-to-Batch Reproducibility of Oasis HLB Sorbent 196

Figure 158: Oasis 2x4 Method 198

Figure 159: Oasis Sorbent Selection 96-Well Plate: Evaluating Oasis 2x4 Method for Cephalexin 199

LIST OF TABLES

Table 1: Comparison of Normal-Phase and Reversed-Phase Chromatography 34

Table 2: Retention Time 43

Table 3: Ion-Exchange Guidelines 46

Table 4: Oasis Family of Ion Exchangers 48

Table 5: Steps Requiring Flow Rate Control 65

Table 6: Relationship Between Cartridge ID and Linear Velocity 67

Table 7: Matrix Effects Scenarios 73

Table 8: Requirements for Condition and Equilibration Steps 102

Table 9: Chromatographic Mode Selection with Sorbents 138

Table 10: Scaling Information 163

Table 11: Scaling for New Conditions 163

Table 12: Scaling Information 163

Table 13: Scaling for New Conditions 164

Table 14: Problems Encountered in the Method Development Process 166

Table 15: Problems Encountered in Executing Existing methods 167

Table 16: Capacity and Elution Volume of Oasis 96-Well Plates and Cartridges 197

Table 17: Tips for Selecting Elution Solvents for the Generic SPE Method [1-D]
The Elution Solvent is Selected Based on Polarity of Analyte 197

Benefits of Solid-Phase Extraction in Sample Preparation

A POWERFUL TOOL FOR IMPROVED SAMPLE PREPARATION

As an analytical scientist, you are faced with many challenges when determining what tools you can best use to achieve the desired result. Determining which sample preparation tools and approaches are important considerations that can significantly impact your success.

Ideally, you would be happy if you did not have to do any sample preparation. In reality, however, sample preparation is often necessary. You may need to optimize a method for an existing sample to improve throughput or lower the cost per analysis. Or, you may be asked to analyze a wide variety of different types of samples to report on new compounds of interest. Each new sample type can present different analytical challenges. In addition, scientists today are faced with the significant challenge of reporting values at lower concentration levels than ever before, without compromising accuracy and precision.

This book is designed to help you explore and understand a very powerful tool in sample preparation technology: solid-phase extraction [SPE]. You will see how this technology, which uses devices with chromatographic packing material, can help meet your analytical challenges.

Definiton of Solid-Phase Extraction

SPE is a sample preparation technology that uses solid particle, chromatographic packing material, usually contained in a cartridge type device, to chemically separate the different components of a sample. Samples are nearly always in the liquid state [although specialty applications may be run with some samples in the gas phase]. Figure 1 shows a sample, which appears black, being processed on a SPE device so that the individual dye compounds, which make up the sample, are chromatographically separated.

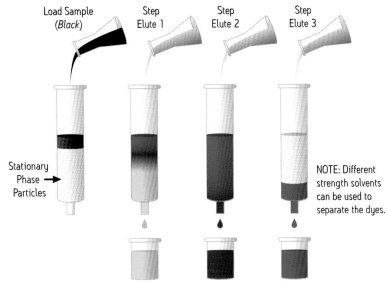

Figure 1: Examples of an SPE Method

The chromatographic bed can be used to separate the different compounds in a sample, to make subsequent analytical testing more successful. For example, SPE is often used for the selective removal of interferences.

The technically correct name for this technology is "Liquid-Solid Phase Extraction," since the chromatographic particles are solid and the sample is in the liquid state. The same basic chromatographic principles of liquid chromatography that are used in HPLC are also used here, but in a different format and for a different reason. Here, chromatography is used to better prepare a sample before it is submitted for analytical testing.

In sample preparation, samples can come from a wide range of sources. They can be biological fluids such as plasma, saliva, or urine; environmental samples such as water, air, or soil; food products such as grains, meat, and seafood; pharmaceuticals; nutraceuticals; beverages; or industrial products. Even mosquito heads can be the sample! When a scientist needed to analyze neuropeptides extracted from the brains of mosquitoes, SPE was the sample preparation method of choice [Waters Applications Database, 1983].

Four Major Benefits of SPE

There are many benefits to using SPE, but four major benefits deserve special attention.

1. Simplification of Complex Sample Matrix along with Compound Purification

One of the most difficult problems for an analytical chemist is when compounds of interest are contained in a complex sample matrix, such as mycotoxins in grains, antibiotic residues in shrimp, or drug metabolites in plasma, serum, or urine. The large number of interfering constituents or substances in the sample matrix along with the compounds of interest makes analysis extremely difficult.

[Four Major Benefits of SPE]

The first problem to solve is the resulting complexity of the analysis itself due to the presence of so many entities which must be separated in order to identify and quantitate the compound[s] of interest. See Figure 2.

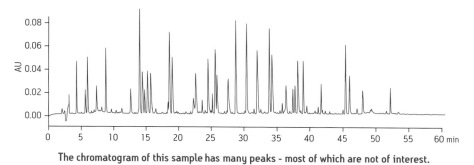

The chromatogram of this sample has many peaks - most of which are not of interest.

Figure 2: Example of a Complex Sample

The robustness of the assay may not be adequate, because any slight change could impact the resolution of the separation of a critical pair of analytes.

Another consideration is that the presence of all the interferences in the original sample matrix can result in instrument downtime due to a buildup of contamination with each injection. If the interferences were removed as part of the sample preparation, then the compounds of interest could be analyzed with a simpler, more robust method. This can be seen in Figure 3, which compares the original sample on top to the new SPE-prepared sample on the bottom.

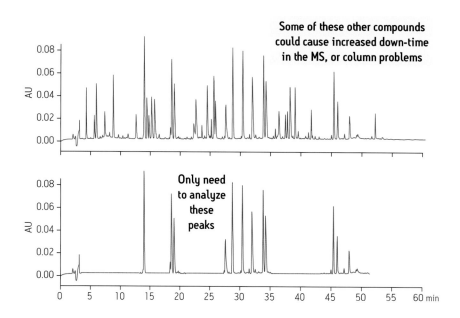

Figure 3: Comparison of Sample Matrix Complexities

An additional benefit of simplifying the sample matrix is improved quantitation accuracy. The top blue trace for compound 1 in Figure 4, initially appears to be acceptable. However, it really has some contamination from the sample matrix when compared to the blank sample matrix trace in red shown just beneath it. With a proper SPE protocol, the lower traces show the same compounds with no problems with interference, making quantitation much more accurate.

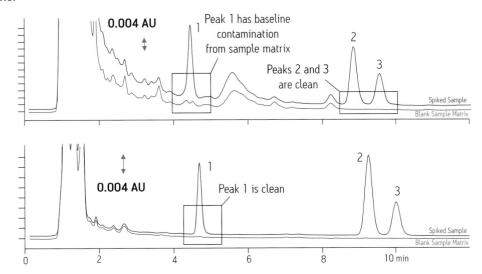

Figure 4: Improved Quantitation with Better Sample Preparation

Another example is shown in Figure 5. The upper trace shows significant interference from the sample matrix on both compounds 1 and 2. The lower trace shows much improved results [clean base line] due to proper sample preparation with SPE. Notice a much cleaner baseline improves the accuracy of the analytical results. Also, a much purer extract can be obtained if the sample requires isolation and purification of that compound.

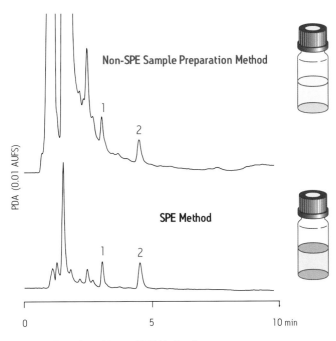

Figure 5: Significant Improvement in Baseline Using SPE Technology

2. Reduce Ion Suppression or Enhancement in MS Applications

The second problem with complex sample matrices can be seen when we look at mass spectrometer output [LC/MS or LC/MS/MS]. For proper MS signal response [sensitivity], the compound ion must be allowed to form properly. In cases where the formation of the compound ion is suppressed by interferences in the sample matrix, the signal strength is greatly diminished.

We can see this effect in Figure 6. The upper output is the signal for our compounds of interest when injected in a saline solution. The lower trace shows significant reduction in response [> 90% suppression] of these same compounds when they were analyzed in human plasma. For the lower trace, only a common protein precipitation step was performed. This technique does not clean up the matrix interferences that cause the ion suppression, resulting in poor signal response.

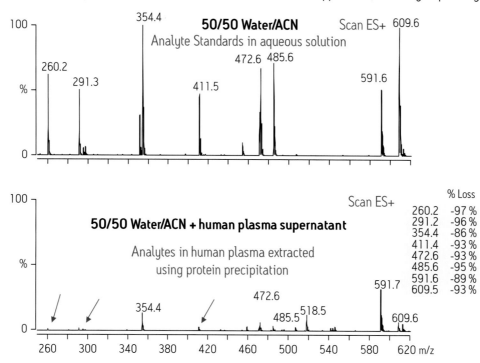

Figure 6: Example of Ion Suppression Due to Sample Matrix

Another good example of this suppression effect can be seen in Figure 7. In the upper trace of the MS output, where the plasma sample was prepared with just a protein precipitation step, we can see that the terfenadine peak is suppressed by 80%. In the lower trace, where the same sample was prepared with a SPE method, we can see minimal ion suppression. Because the interferences from the sample matrix were removed, this allowed the compound ion to form properly, creating a better signal.

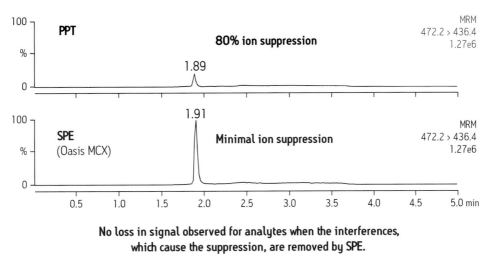

Figure 7: Reduced Ion Suppression with Proper SPE

In some instances, interferences from the sample matrix can artificially increase the signal reported for a compound. This is called ion enhancement, resulting in an inaccurately high reported value. A proper SPE method will minimize this effect by cleaning away the interferences from the compound, resulting in a more accurate reported value.

3. Capability to Fractionate Sample Matrix to Analyze Compounds by Class

An analyst may be faced with a sample that contains many compounds, with a need to separate them by class so that further analysis can be carried out much more efficiently. For example, a soft drink beverage contains a wide range of compounds in its formulation. An SPE method could be developed to separate the different classes of compounds, for example by their polarity. The polar compounds could be collected, as a separated fraction, from the more non-polar compounds. These two fractions could then be separately analyzed in a much more efficient way because their compounds would be more similar.

[Four Major Benefits of SPE]

An example of the power of fractionation by SPE is shown in Figure 8. Here, a complex sample of a dry powder [purple grape drink mix] is easily separated into four fractions: a fraction of just the polar compounds, a purified red compound, a purified blue compound, and a fraction containing all the remaining very non-polar compounds. You will see elsewhere in this book how very powerful this capability can be.

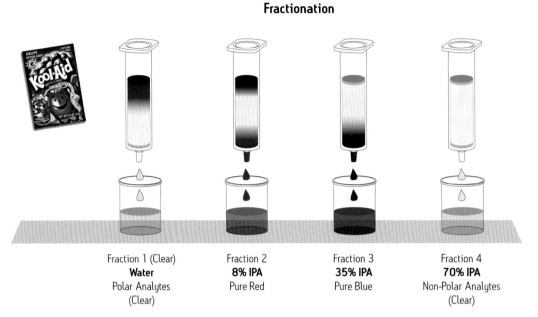

Figure 8: Sample Preparation by SPE

For a more detailed discussion of sample fraction by SPE see page 125 in the Method Development section.

[Four Major Benefits of SPE]

4. Trace Concentration [Enrichment] of Very Low Level Compounds

Analysts today often need to report on compounds at far lower concentration levels than ever before, as little as parts per trillion [ppt] and even lower. Typically these levels are lower in the neat sample than the sensitivity capability of the analytical instruments.

A good example of this is the analysis for trace contaminants in environmental samples or metabolite development over time in biological fluids. The upper trace in Figure 9 shows the poor response of the original neat sample for the compound of interest. Using the same analytical conditions but with the sample prepared with SPE used in a trace concentration strategy, the lower trace shows a dramatic increase in signal strength for this compound. With this result, an accurate calculation of the original compound concentration in the neat sample can be made.

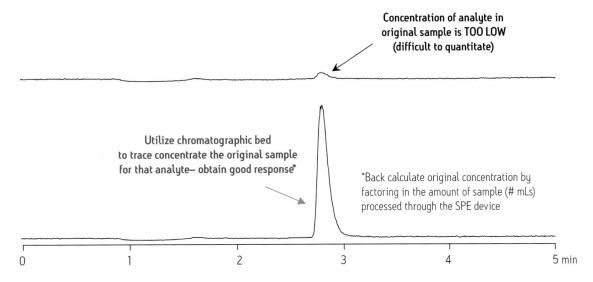

Figure 9: Example of Trace Concentration

Without the retention capability of chromatographic packing materials in SPE, the ability to trace concentrate a specific compound[s] would be very difficult, if not impossible, with other sample preparation approaches.

[Four Major Benefits of SPE]

Summary

As we've seen, an SPE device with a chromatographic bed can perform four critical functions to make the analysis of the sample more successful. See Figure 10.

The chromatographic bed in the cartridge can perform four critical functions:

1. Sample simplification
2. Matrix effects reduction
3. Fractionation
4. Trace concentration

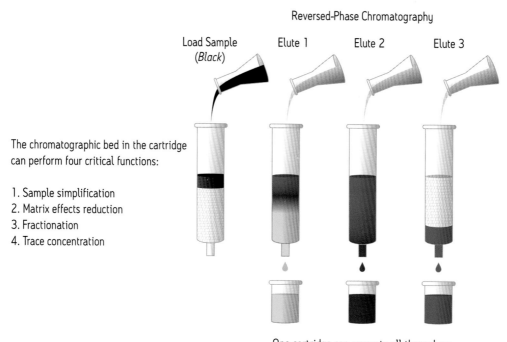

Figure 10: The Power of SPE

In this book, we have endeavored to provide all of the SPE fundamentals and success techniques derived from scientists from all over the world who have counted on this technology in the past thirty years. Today, scientists are finding SPE more useful than ever in solving difficult sample preparation and analytical problems.

We hope this book will enable you to understand and master the capabilities of SPE, so that you too can put the power of this technology to use in your laboratory.

SPE is LC

MODES OF LIQUID CHROMATOGRAPHY

The power of SPE to solve difficult sample preparation problems is created by the use of liquid chromatography [LC] to perform chemical separations on the sample. Whether the goal is sample matrix simplification, fractionation, or trace concentration [enrichment], the analytical scientist can create the proper chromatographic conditions by manipulating the choice of sorbent [stationary phase], and solvents [mobile phase]. The chromatographic principles are the same as those in HPLC and UPLC®, which are familiar to you. For more information on LC Technology, see page 208 in the Appendix.

Based on the goal of the SPE protocol, the first step is to identify what is known about the sample matrix, the compounds of interest, and the subsequent analytical technology. Then the mode of chromatography and SPE strategy can be chosen.

As in LC, several options are available:

- Normal phase
- Reversed phase
- Ion exchange
- Affinity

And, specifically for SPE:

- Specialty techniques
- Solid-phase reaction chemistry

This section will describe the main principles of LC and how to create the power of chemical separation in the SPE protocol.

Chemical Separation Power—Selectivity

The choice of a combination of particle chemistry [stationary phase] and mobile-phase composition—the separation system—will determine the degree of chemical separation power [how we change the speed of each analyte]. Optimizing selectivity is the most powerful means of creating a separation; this may reduce the need for the brute force of the highest possible mechanical efficiency. To create a separation of any two specified compounds, a scientist may choose among a multiplicity of phase combinations [stationary phase and mobile phase] and retention mechanisms [modes of chromatography]. These are discussed in the next section.

LC Separation Modes

In general, three primary characteristics of chemical compounds can be used to create LC separations. They are:

- Polarity
- Electrical Charge
- *Molecular Size (not covered in this book)*

[Separations Based on Polarity]

First, let's consider polarity and the two primary separation modes that exploit this characteristic: normal-phase and reversed-phase chromatography.

Separation Based on Polarity

How It Works

This approach takes advantage of the polarity or non-polarity of the subject compound. As we know, a molecule's structure, activity, and physicochemical characteristics are determined by the arrangement of its constituent atoms and the bonds between them. Within a molecule is one or more functional groups, a specific arrangement of certain atoms that is responsible for special properties and predictable chemical reactions. It is this structure that often determines whether the molecule is polar or non-polar. We can see in Figure 11 that water [a small molecule with a high dipole moment] is a very polar compound.

As shown in Figure 11, classes of molecules can be ordered by their relative retention into a range or spectrum of chromatographic polarity from highly polar to highly non-polar.

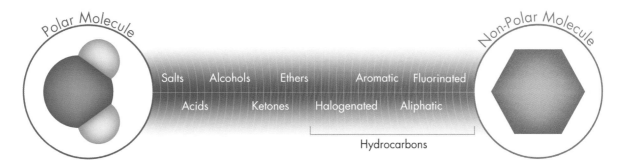

Figure 11: Polarity Spectrum

Benzene [an aromatic hydrocarbon] is a non-polar compound. Molecules with similar chromatographic polarity tend to be attracted to each other; those with dissimilar polarity exhibit much weaker attraction, if any, and may even repel one another. This becomes the basis for chromatographic separation modes based on polarity. Another way to think of this is by the familiar analogy: oil [non-polar] and water [polar] don't mix. Unlike in magnetism, where opposite poles attract, chromatographic separations based on polarity depend upon the stronger attraction between likes and the weaker attraction between opposites.

Remember, "like attracts like" in polarity-based chromatography.

[Separations Based on Polarity]

Designing a Polarity–Based Method

To design a chromatographic separation system [see Figure 12], we create competition for the various compounds contained in the sample by choosing a mobile phase and a stationary phase with different polarities.

Figure 12: Proper Combination of Mobile and Stationary Phases Affects Separation Based on Polarity

Then, compounds in the sample that are similar in polarity to the stationary phase [column packing material] will be delayed because they are more strongly attracted to the particles. Compounds whose polarity is similar to that of the mobile phase will be preferentially attracted to it and move faster. Based upon differences in the relative attraction of each compound for each phase, a separation is created by changing the speeds of the analytes.

Figures 13, 14, and 15 display typical chromatographic polarity ranges for mobile phases, stationary phases, and sample analytes, respectively. Let's consider each in turn to see how a chromatographer chooses the appropriate phases to develop the attraction competition needed to achieve a polarity-based LC separation.

Mobile Phase

A scale, such as that shown in Figure 13, upon which some common solvents are placed in order of relative chromatographic polarity is called an eluotropic series.

Figure 13: Mobile Phase Chromatographic Polarity Spectrum

Mobile phase molecules that compete effectively with analyte molecules for the attractive stationary phase sites displace these analytes, causing them to move faster through the column [weakly retained]. Water is at the polar end of the mobile-phase-solvent scale, while hexane, an aliphatic hydrocarbon, is at the non-polar end. In between, single solvents, and miscible-solvent mixtures [blended in proportions appropriate to meet specific separation requirements] can be placed in order of elution strength. Which end of the scale represents the 'strongest' mobile phase depends upon the nature of the stationary phase surface where the competition for the analyte molecules occurs. For example, in normal-phase application with a polar sorbent, water is a strong solvent because it is also polar and can release analytes from the sorbent. If it is a reversed-phase application with a non-polar sorbent, then water is a weak solvent, since it cannot release non-polar analytes that are attracted to the sorbent.

Stationary Phase

Unbonded silica has an active, hydrophilic [water-loving] polar surface containing acidic silanol [silicon-containing analog of alcohol] functional groups. Consequently, it falls at the polar end of the stationary-phase scale shown in Figure 14.

Figure 14: Stationary-Phase Particle Chromatographic Polarity Spectrum

The activity or polarity of the silica surface may be modified selectively by chemically bonding to it less polar functional groups [bonded phase]. Examples shown here include, in order of decreasing polarity, cyanopropylsilyl- [CN], n-octylsilyl- [C_8], and n-octadecylsilyl- [C_{18}, ODS] moieties on silica. The latter is a hydrophobic [water-hating], very non-polar packing.

Analyte Characteristics

Figure 15 shows the chromatographic polarity spectrum of our sample analytes.

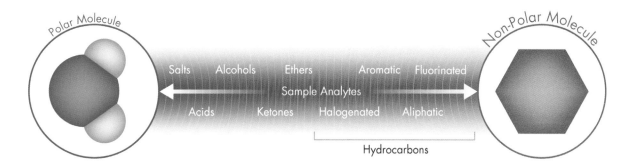

Figure 15: Compound/Analyte Chromatographic Polarity Spectrum

For example, some compounds are very polar, such as salts and charged acids, while other analytes, are very non-polar such as flavoring oils.

[Normal-Phase SPE]

After considering the polarity of both phases, a chromatographer must choose a mobile phase for a given stationary phase in which the analytes of interest are retained, but not so strongly that they cannot be eluted. Among solvents of similar strength, the chromatographer considers which phase combination may best exploit the more subtle differences in analyte polarity and solubility to maximize the selectivity of the chromatographic system. Like attracts like, but creating a separation based upon polarity involves knowledge of the sample and experience with various kinds of analytes and retention modes.

To summarize, the chromatographer should choose the best combination of a mobile phase and stationary phase particle with appropriately opposite polarities. Then, as the sample analytes move through the column, the rule "like attracts like" will determine which analytes slow down and which proceed at a faster speed to elute first.

Normal-Phase SPE

In his separations of plant extracts, Dr. Tswett, the father of chromatography, was successful using a polar stationary phase [chalk or alumina in a glass column] with a much less polar [non-polar] mobile phase. These conditions are now known as the "normal-phase mode of chromatography." In fact, it appears that the origins of today's liquid chromatography were really an early SPE experiment to clean up a complex sample matrix!

Figure 16 represents a normal-phase chromatographic separation of a three-dye test mixture.

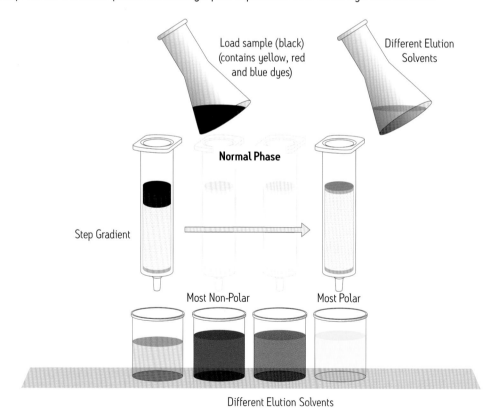

Figure 16: Normal-Phase Chromatography

In normal-phase chromatography, the stationary phase is polar and retains the polar yellow dye most strongly. The relatively non-polar blue dye is more attracted by the mobile phase, a non-polar solvent, and elutes quickly. Since the blue dye is most like the mobile phase [both are non-polar], it moves faster and elutes first. The mobile phase used for normal-phase chromatography on silica is 100% organic; no water is used.

Reversed-Phase SPE

The term reversed phase describes the chromatography mode that is just the opposite of normal phase, namely the use of a polar mobile phase and a non-polar [hydrophobic] stationary phase. Figure 17 illustrates the three-dye mixture being separated using such a protocol. Note the similarity between SPE and LC.

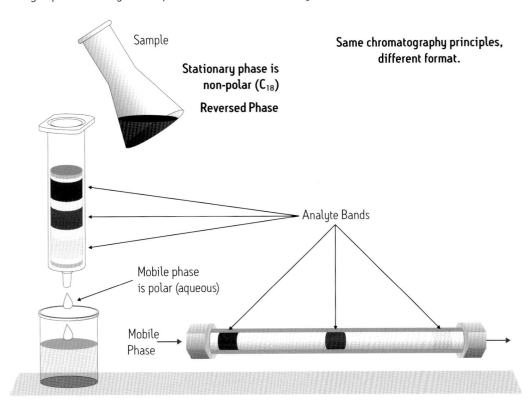

Figure 17: Reversed-Phase Chromatography, SPE and LC

The most strongly retained compound in this example is the more non-polar blue dye, as its attraction to the non-polar stationary phase is greatest. The polar yellow dye, being weakly retained, is attracted by the polar, aqueous mobile phase and moves the fastest through the bed and elutes earliest. Remember, like attracts like.

[Reversed-Phase SPE]

Today, because it is more reproducible and has broad applicability, reversed-phase chromatography is used for over 60% of all SPE and LC methods. Most of these protocols use a mobile phase comprised of an aqueous blend of water with a miscible, polar organic solvent, such as acetonitrile or methanol. This typically ensures the proper interaction of analytes with the non-polar, hydrophobic particle surface. A C_{18}—bonded silica [sometimes called ODS] is the most popular type of reversed-phase packing material or sorbent. [Figure 18]

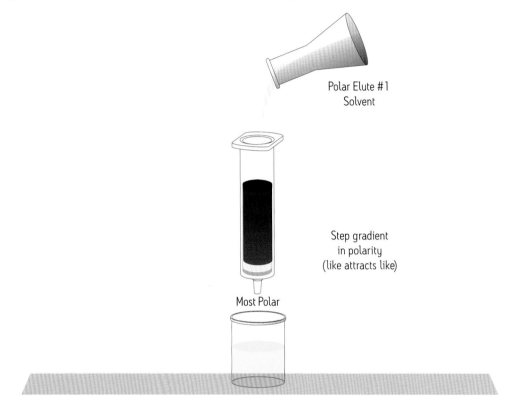

Figure 18: Reversed-Phase SPE

Table 1 below summarizes the phase characteristics and analyte behavior for the two principal LC/SPE separation modes based upon polarity. Remember, for these polarity-based modes, like attracts like.

Table 1: Comparison of Normal-Phase and Reversed-Phase Chromatography

	Normal Phase	Reversed Phase
Sorbent	Polar	Non-Polar
Mobile Phase [Initial Conditions]	Non-Polar	Polar
Analytes Most Strongly Retained	Polar	Non-Polar
Analytes Eluting First	Non-Polar	Polar

Importance of Conditioning and Equilibration Steps

This topic discusses why they are so important, when reversed-phase SPE conditions are used. In reversed-phase chromatography, the porous sorbent particles are very non-polar, and analytes are typically in aqueous, more polar solvents. Because the sorbent and mobile phase solvent are opposites in polarity, there is no attraction between them. A good analogy of this situation is a comparison of an old, poorly maintained, and oxidized automobile surface [very polar] and a freshly waxed surface [very non-polar]. Focus on the behavior of water on these two surfaces. [Figures 19 and 20]

Unbonded silica particles are very polar and would attract a polar mobile phase, such as water. This is similar to what happens to water on the surface of an old automobile with a very oxidized paint surface, which is very polar.

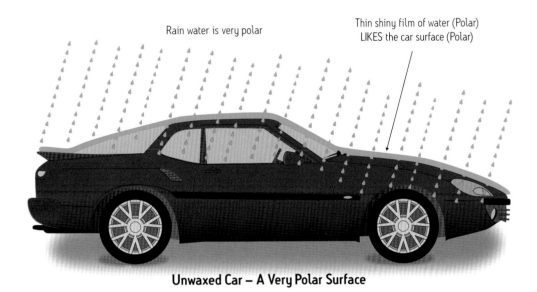

Figure 19: Polar Surface with Polar Liquid

The water molecules are attracted to the polar surface and form a thin film.

In the discussion of reversed-phase SPE methods, we need to understand the importance of the conditioning and equilibrium steps in the SPE protocol. [Addtional information on these 2 steps is provided starting on page 102]

[Importance of Conditioning and Equilibration Steps]

When you "wax" your car, you are applying a very non-polar surface, see Figure 20.

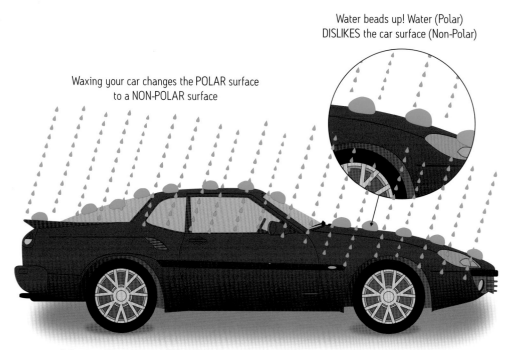

Waxing Your Car is Like Bonding C_{18} to Silica – it Makes it Non-Polar

Figure 20: Non-Polar Surface and Polar Liquid

The polar water molecules do not want to be near the non-polar wax surface. They would prefer to be with other polar water molecules, and with surface tension, they form "beads" of polar water, which can then roll off the non-polar surface.

Reversed-phase SPE sorbents are very porous, which provide the high surface area needed to perform the chromatographic separation with a relatively small amount of sorbent mass. When bonded with carbon ligands such as C_{18}, the pores are very non-polar. To access this chromatographic surface area, analytes [compounds of interest] that are in the sample must travel into the pores filled with solvent so they will be attracted and retained. See Figure 21. The amount of time that the analyte spends entering and then eventually exiting the pore is what determines the amount of retention that the SPE cartridge generates.

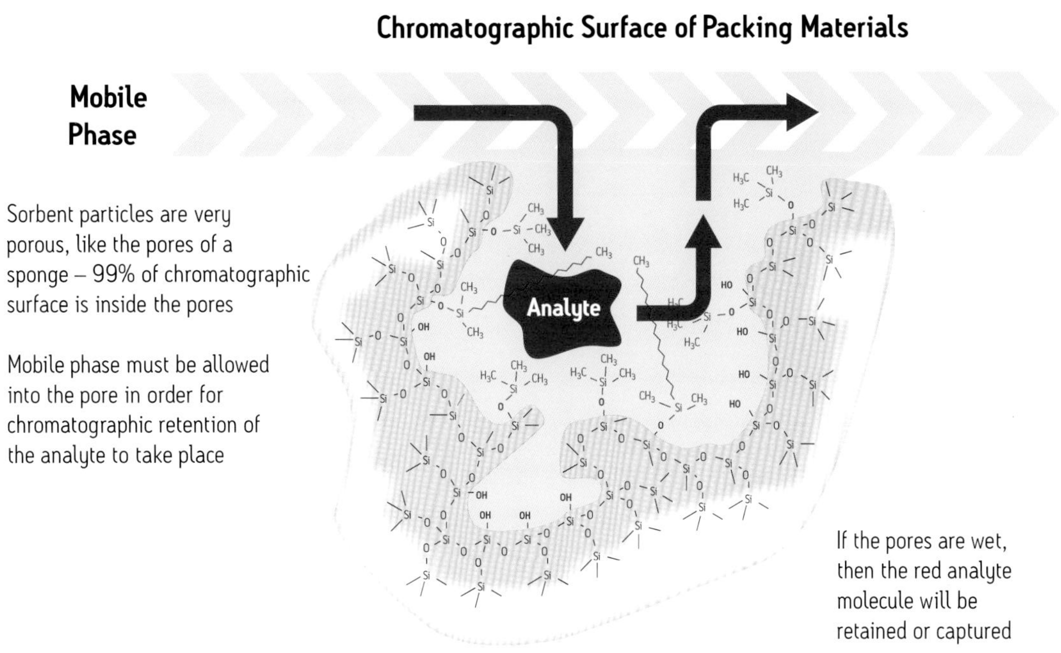

[Importance of Conditioning and Equilibration Steps]

Chromatographic Surface of Packing Materials

Mobile Phase

Sorbent particles are very porous, like the pores of a sponge – 99% of chromatographic surface is inside the pores

Mobile phase must be allowed into the pore in order for chromatographic retention of the analyte to take place

If the pores are wet, then the red analyte molecule will be retained or captured

Figure 21: Reversed-Phase Sorbent with Wetted Pore

The conditioning and equilibration steps ensure that the pores of a reversed-phase sorbent are properly wetted to allow the analyte to be retained. However, if the polar sample solvent does not want to be in the non-polar pores, because of opposite polarity, then the analytes cannot travel into the pores and the chromatographic retention of the compounds does not occur. The SPE protocol will stop working.

Figure 22 shows this effect. Note that the pore on the right actually dries out. This is called "de-wetting." This behavior has been misunderstood in the past and referred to as hydrophobic collapse. However, it is actually caused by the pores not being properly wetted. Conditioning and equilibration steps are required to properly wet the pores of a non-polar sorbent. These steps will be discussed in the next few pages.

[**Importance of Conditioning and Equilibration Steps**]

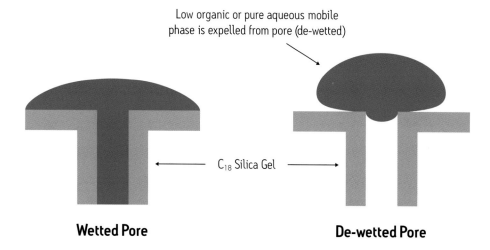

Note: Retentivity is a function of the surface area and ligand density. However, if the surface is non-wetted, then the effective chromatographic surface area is reduced > 95%, therefore reducing the retentivity of the analyte. This results in poor capture and appears as if hydrophobic-phase collapse has occured. Remember that almost all of the surface area is in the pores.

Figure 22: Pore De-Wetting

If the sorbent pores are de-wetted, then the analyte, which is supposed to enter the pore and be attracted and captured by the pore surface of the sorbent, cannot get into the pore. As seen in Figure 23, the analyte quickly elutes from the cartridge.

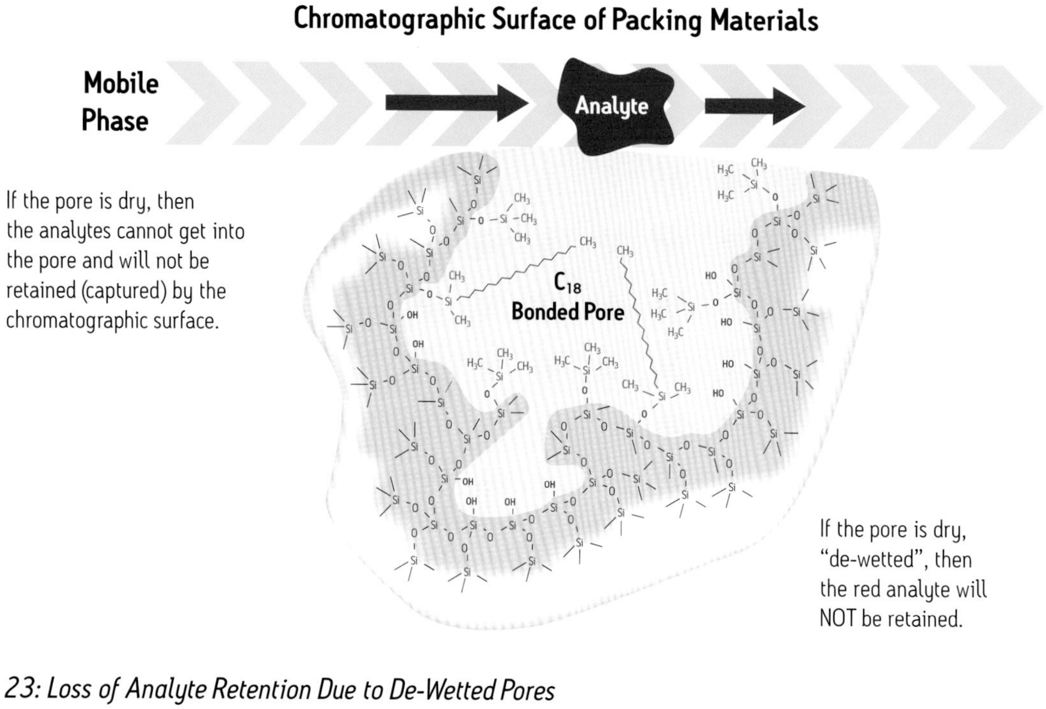

Chromatographic Surface of Packing Materials

Mobile Phase → Analyte →

If the pore is dry, then the analytes cannot get into the pore and will not be retained (captured) by the chromatographic surface.

C_{18} Bonded Pore

If the pore is dry, "de-wetted", then the red analyte will NOT be retained.

Figure 23: Loss of Analyte Retention Due to De-Wetted Pores

Effectively, the conditioning step uses a solvent that is attracted into the reversed-phase, non-polar pores, properly wetting them. Methanol and acetonitrile are commonly used. These solvents are typically too strong to allow the compounds in the sample to be retained. So they must be removed before the sample is loaded. A different equilibration solvent, which is miscible with the conditioning solvent but not as strong chromatographically, now replaces the conditioning solvent in the pores. This equilibration solvent will typically be the same solvent that is in the sample. Water can also be used as the equilibration solvent because it is miscible with the methanol and acetonitrile, and they bring the water into the pores even though the pores are non-polar. The sample can then be loaded, and the compounds can travel into the wetted pores and be retained. See additional discussion on page 102.

Therefore, for reversed-phase SPE, the conditioning and equilibration steps are critical, Figure 24.

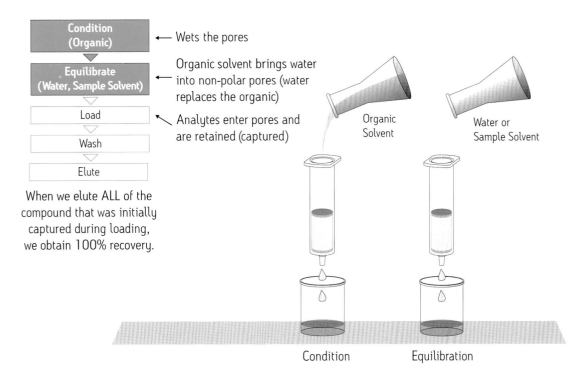

Figure 24: Conditioning and Equilibration Steps

The Effect of De-Wetting Phenomena/Drying Out Effect on SPE

An important point to remember is that in the day-to-day execution of a reversed-phase SPE protocol, especially with a C_{18} or C_8 silica-based sorbent particle using a vacuum manifold, major recovery problems can occur from sample-to-sample related to de-wetting of sorbent pores.

If care is not taken during the conditioning and equilibration steps, especially when using a vacuum manifold, the pores of the sorbent may not be wet in some of the SPE cartridges, causing poor retention of the compounds, and thus poor recovery for samples processed in those cartridges.

[The Effect of De-Wetting Phenomena/Drying Out Effect on SPE]

The problem occurs if multiple cartridges or multiple wells in a 96-well plate device are being conditioned simultaneously on a vacuum manifold. See Figure 24 on page 39 and Figure 91 on page 95. Typically, the conditioning solvent is added to all the cartridges [wells], and then the vacuum is applied to the manifold to pull the conditioning solvent down through the sorbent bed.

The conditioning solvent wets all the pores of the sorbent, and brings the equilibration solvent into the pores as described previously. However, the flow of each cartridge or well during the conditioning step can be different. If a cartridge or well flows too fast, then all of the conditioning solvent can be pulled through that cartridge before the other cartridges have been completed. Since the vacuum is still on for the slower cartridges, air will be drawn into the sorbent bed of the faster cartridges. The air will evaporate the conditioning solvent from the pores, and the longer the vacuum is on, the more the pores will dry out. [Figure 23]

Further, when the equilibration step is performed, the pores in the cartridges and wells that dried out have little or no conditioning solvent present to bring the equilibration solvent into those pores. Therefore, those pores remain dry. Any sample compounds that are supposed to be retained by traveling into wetted pores are therefore not captured. This eventually results in a poor recovery percentage for those samples. The image on the left of Figure 25 shows a properly wetted cartridge, all 10 ng of the pure green analyte is captured in the load step. However, in a partially de-wetted cartridge [image on the right], 4 ng of the green analyte passes right through and is not captured, leaving only 6 ng on the cartridge which results in a very poor 60% recovery. The sample preparation steps will have to be repeated!

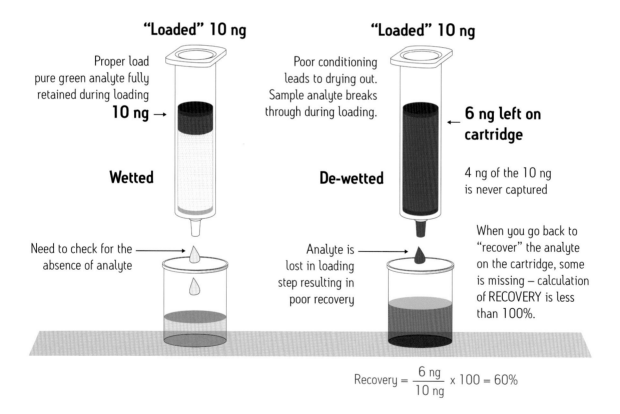

Figure 25: Poor % Recovery Due to De-Wetting

On a given day, there is the potential that samples may have to be retested due to this problem. Some vacuum manifolds use valves [stop-cock valves] to try to control this problem; however, it is still difficult to keep all the cartridges wet. Processing 96-well plates is even more difficult, since there is no way to control the flow in individual wells.

A solution to this drying out problem is to use sorbent particles developed specifically for this purpose. Oasis HLB sorbent is a reversed-phase particle made from a wettable copolymer that will resist this drying out effect resulting in greater reproducibility in recovery performance every day. The Oasis HLB sorbent is a copolymer of divinylbenzene and n-vinylpyrrolidone. This SPE sorbent provides, a balance of hydrophilic and lipophilic [HLB] properties which are ideal for reversed-phase SPE. [Technical information on this Oasis sorbent is included in the Appendix on page 192.]

Retention Mechanism Map for Reversed-Phase SPE

Since reversed-phase chromatography is the most popular mode of SPE, it is important to understand how best to control the conditions in the cartridge to provide the most powerful sample preparation possible. A sample will contain a wide variety of compounds and substances, and could have both polar and non-polar constituents that could be acids, bases, or neutral molecules. Under reversed-phase conditions, the sorbent will be non-polar and will attract, or retain [capture], the non-polar molecules. These can then be selectively eluted by using stronger elution solvents.

However, an important tool in SPE methods development is to use pH as a part of the solvent selection and separation processes. The significance of this is seen in the following retention map for a pure reversed-phase SPE protocol. See Figure 26. This map plots the behavior of the compounds under "pure" reversed-phase conditions. No ion-exchange mixed mode sorbent behavior is present.

The retention map shows the relationship of the k for a compound [retention or capture] on the Y-axis, and the solvent pH on the X-axis, with low to high pH from left to right. The pH will refer to the aqueous portion of the mobile phase being used in the method. The mobile phase may be a mixture of organic solvent with this aqueous component.

It is important to remember what type of reversed-phase sorbent particle is contained in the cartridge, when pH is being adjusted. For silica-based particles [C_{18} etc.] the pH should not be allowed to go higher that 8, because the silica particle will begin to dissolve, possibly contaminating the sample. The Oasis HLB reversed-phase particle is a copolymer, so pH does not dissolve the particle. This provides a wider range of method development conditions to create the most successful SPE protocol. For a given map, the percent organic in the solvent system is held constant, and only the pH is adjusted. [Figure 26]

[Retention Mechanism Map for Reversed-Phase SPE]

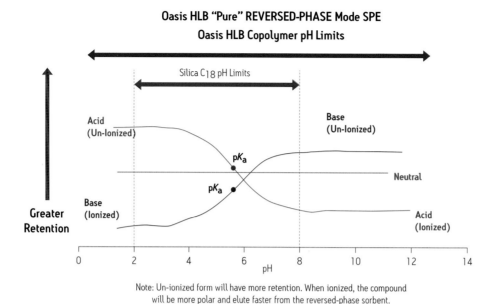

Figure 26: Retention Map for Oasis HLB Pure Reversed Phase

The three types of compounds are plotted: neutral compounds [cannot become ionized], acids, and bases. Acids and bases can be either un-ionized or ionized depending on the pH of the solvent. Note the very different retention behavior of the three compound types.

Neutral Compounds

First, a neutral compound sees no change in its retention as the pH changes from low to high, since it cannot become ionized. Its k value will be determined by its hydrophobicity, the hydrophobicity of the sorbent, and the strength of the organic solvent modifier. Changes in pH have no effect on its retention.

Acids

An acid shows a significant change in its k [retention or capture] when only the pH is adjusted. Note that at low pH, on the left, the acid is in its un-ionized form. The retention will be very high, with a typical k > 10. As the pH is increased, moving to the right, the acid becomes more and more ionized. The retention in its ionized form is significantly less, with k = ~1. The retention curve for a compound will pass through its half way point when the compound is at its pKa point in that solvent.

Bases

A base also shows a significant change in retention. At low pH, on the left, the base is in its ionized form with very low retention, k = ~1. However, as the pH is adjusted upward, the base becomes un-ionized, and the retention goes up significantly, with a much higher k value.

Interpreting the map shows how powerful it can be in developing SPE protocols. The way to predict the elution order of the compounds is to pick a pH value, on the X-axis, and read up on the Y-axis to see the retention order. Refer back to Figure 26 and map the following 3 different pH values to see the elution order. See Table 2. For example, the elution order at pH 10 is the ionized acid followed by the neutral, and then followed by the un-ionized base.

Table 2: Retention Time

Analyte Retention	pH 2	pH 6	pH 10
Greatest	Un-ionized Acid		Un-ionized Base
Moderate	Neutral	Neutral	Neutral
Lowest	Ionized Base	Co-Elution Acid/Base	Ionized Acid
	pH 2	pH 6	pH 10

Changing the pH will totally change the selectivity [elution order] of these compounds.

The Power of pH

The rule to remember is that in a reversed-phase system, the ionized form of a compound will have little retention and the un-ionized form will show significantly more retention. In most cases, intermediate pH values should be avoided if possible, because the steep slope for the curves in that region indicates that slight variations in the pH will cause large changes in retention, which could result in poor SPE performance due to reproducibility problems. In addition, the pka ± 2 rule will also be used. This predicts the charge state as a function of the pH. A compound with a pka = 5 will either be fully charged or completely uncharged in a pH range of 3–7. For example, an acid with a pka = 5 will be fully charged at pH = 7 and completely uncharged at pH = 3.

For SPE method development, this relationship with pH can be very powerful. For example, if the pass-through strategy is to be used, the compound should be kept in its ionized form in the loading and elution steps for very low retention so it can pass through easily because of its low k value.

If the capture strategy* is used, then the compound should be kept in its un-ionized form [k = high] for the load and wash steps to ensure maximum capture, and then adjust the pH to create the ionized form [k = low] for much less retention during the elution step. In trace enrichment/concentration methods, adjust the pH to keep the compound in its un-ionized form [k = high] to ensure complete capture in the load step.

The reversed-phase retention map shows the power of adjusting the pH of the solvent to reach the goals of the SPE protocol. Remember, silica-based sorbents should not be exposed to pH values greater than 8, or they will begin to dissolve. Also, silica-based reversed-phase sorbents can cause poor recoveries for bases if the pH is not properly controlled due to the creation of an unexpected cation exchange mechanism. See the discussion on page 168 in the Troubleshooting section. The Oasis HLB copolymer does not have that limitation. It will provide a greater operating range for method development and has the added benefit of not drying out, as mentioned before.

* The different SPE strategies are discussed beginning on page 116.

[Separations Based on Charge—IEC]

Separations Based on Charge: Ion-Exchange Chromatography [IEC]

For separations based on polarity, like attracts like and opposites may repel. In ion-exchange chromatography, and other separations based upon electrical charge, the rule is reversed. Likes may repel, while opposites are strongly attracted. For an ion-exchange retention mechanism to occur, the analyte and the sorbent must be oppositely charged.

Stationary phases for ion-exchange separations are characterized by the charge on their surfaces and the types of ions that they attract and retain. Anion exchange is used to retain and separate negatively charged ions on a positive surface. Conversely, cation exchange is used to retain and separate positively charged ions on a negative surface. [Figure 27]

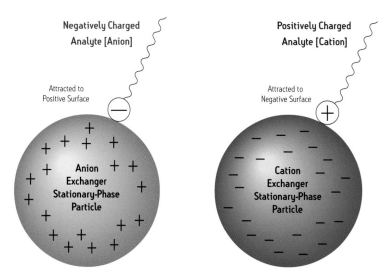

Figure 27: Ion-Exchange Chromatography.

An anion exchanger will have a positive [+] charge on its surface and retain anions that have an opposite negative [–] charge. A cation exchanger will have a negative [–] charge on its surface and retain cations that have an opposite positive [+] charge.

Important Note: To make it easier to understand ion-exchange chromatography, there are two definitions that will be important to remember: strong and weak. The term "strong" will indicate either a compound or a sorbent, which will always be charged during the method. The term "weak" will indicate either a compound or a sorbent, which may be charged or may not be charged [un-ionized] depending on the pH of the solution solvent system used during the method.

Ionization States

Charge-based separations depend on an understanding of the ionization states of the analytes [acids and bases] and the sorbents involved.

Figure 28 will clarify the ionization state and charge for a compound molecule as a function of the pH of the solution.

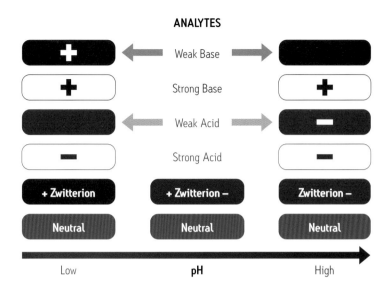

Figure 28: Charged State of Analytes vs. pH

Note the definition of strong [always charged] and weak [on or off] is important so that you can understand why the chromatography behaves in a certain way. Some compounds are called zwitterions because they have both acidic and basic sites, which both can be charged at mid pH values.

The ion exchangers are the sorbent particles that are packed into the SPE devices. A very special family of these sorbents was created specifically for SPE built on that Oasis HLB reversed-phase co-polymer. By functionalizing the particle surface with specific charged sites creates a sorbent with both ion-exchange and reversed-phase properties. Figure 29 shows the various charge states for the Oasis ion exchangers. Refer to the technical information on Oasis sorbents on page 48.

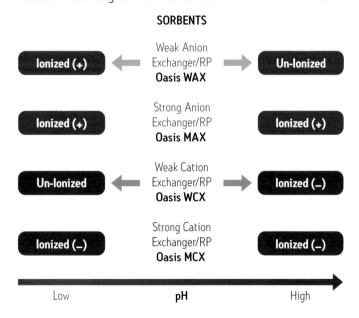

Figure 29: Oasis Family of Ion-Exchange Sorbents [RP = Reversed Phase]

[Separations Based on Charge—IEC]

Strong ion exchangers bear functional groups [e.g., quaternary amines or sulfonic acids] that are always ionized. They are typically used to retain and separate weak ions. These weak ions may be eluted by displacement with a mobile phase containing ions that are more strongly attracted to the stationary phase sites. Alternately, weak ions may be retained on the column, then turned into their un-ionized form in situ by changing the pH of the mobile phase, causing them to lose their attraction and elute. With the advent of LC-MS applications, more and more SPE methods are using ion-exchange combined with reversed-phase modes of chromatography in the same SPE device, since it provides cleaner extract that reduces ion suppression or ion enhancement on the mass spectrometer.

Weak ion exchangers [e.g., with secondary-amine or carboxylic-acid functions] may be neutralized above or below a certain pH value and lose their ability to retain ions by charge. When charged, they are used to retain and separate strong ions. If these ions cannot be eluted by displacement, then the stationary phase exchange sites may be neutralized, shutting off the ionic attraction and permitting elution of the charged analytes.

Table 3 provides guidance on the selection of sorbent type and solution pH for ion-exchange SPE methods. First, determine analyte type. Then, follow corresponding arrows down for recommended particle and mobile phase pH. [In some cases, sorbents may have both ion-exchange and reversed-phase properties.] When weak ion exchangers are put into their un-ionized form, they may retain and separate species by hydrophobic [reversed phase] or hydrophilic [normal phase] interactions; in these cases, elution strength is determined by the polarity of the mobile phase. Thus, weak ion exchangers may be used for mixed-mode separations [separations based on both polarity and charge].

Analyte Type	Weak ACID e.g., $pk_a \sim 5$		Strong Acid	Weak Base e.g., $pk_a \sim 10$		Strong BASE
Charge State vs. pH*	No charge at pH < 3	– [anion] at pH > 7	– [anion] Always Charged	+ [cation] at pH < 8	No charge at pH > 12	+ [cation] Always Charged

Stationary Phase Particle	Strong Anion Exchanger	Weak Anion Exchanger e.g., $pk_a \sim 10$		Strong Cation Exchanger	Weak Cation Exchanger e.g., $pk_a \sim 5$	
Charge State vs. pH*	+ Always Charged	+ at pH < 8	No charge at pH < 12	– Always Charged	No Charge at pH < 3	– at pH > 7
Mobile Phase pH Range						
to Retain analyte [capture]	pH > 7	pH < 8		pH < 8	pH > 7	
to Release analyte [elute]	pH < 3	pH > 12		pH > 12	pH < 3	

*Note: pH Ranges are approximate. The actual values will depend upon specific analyte and particle characteristics

Table 3: Ion-Exchange Guidelines

The table above provides guidelines for the principal categories of ion exchange. For example, to retain a strongly basic analyte [always positively charged], use a weak-cation-exchange stationary phase particle at pH > 7; this assures a negatively charged particle surface. To release or elute the strong base, lower the pH of the mobile phase below 3; this removes the surface charge of the particle and shuts off the ion-exchange retention mechanism. Note that a pKa is the pH value at which 50% of the functional group is ionized and 50% is un-ionized. To assure an essentially unionized, or a fully ionized analyte or particle surface, the pH must be adjusted to a value at least ±2 units beyond the pKa, as appropriate [refer to Table 3].

Important: Do not use a strong-cation exchanger to retain a strong base; both will remain charged and strongly attracted to each other, making the base nearly impossible to elute. It can only be removed by swamping the strong cation exchanger with a competing base that exhibits even stronger retention and displaces the compound of interest by winning the competition for the active exchange sites. This approach is rarely practical, or safe, in LC and SPE. [Very strong acids and bases are dangerous to work with, and they may be corrosive to materials of construction used in LC fluidics!]

For ion-exchange chromatography, it is important to keep the state and charge for the compound and the sorbent in mind so that the chromatographic performance will be predictable. To summarize:

1. The sorbent and compound must have opposite charges to activate the retention mechanism.
2. The term "strong" [compound or sorbent] indicates that it is ALWAYS charged.
3. The term "weak" [compound or sorbent] indicates that it CAN BE charged [or it can be un-ionized].
4. To release the compound from the SPE sorbent bed, one of the charges has to be shut off, either on the sorbent, or the compound, by adjusting the pH.

Sorbents for Strong Compounds

If the compound is "strong" it will ALWAYS be charged; therefore a "weak" sorbent must be used. The sorbent will be opposite charged for capture and then un-ionized for release of the "strong" compound by adjusting the pH.

Sorbents for "Weak" Compounds

If the compound is, "weak" use the "strong" sorbents, which are ALWAYS charged. The compound will be opposite charged for capture, and then un-ionized for release by adjusting the pH.

[Strong Cation-Exchange and Reversed-Phase Sorbent]

Choosing an Ion Exchange Sorbent for SPE

The reversed-phase Oasis HLB copolymer particles, can be made into four different ion-exchange sorbents with the same co-polymer backbone. See Table 4 and page 194.

Table 4: Oasis Family of Ion Exchangers

	Anion Exchanger	Cation Exchanger
Strong	Oasis MAX	Oasis MCX
Weak	Oasis WAX	Oasis WCX

These ion-exchange sorbents are functionalized from the reversed-phase Oasis HLB particles, and incorporate both the ion-exchange and reversed-phase retention mechanisms, making them even more powerful for solving difficult SPE application problems. Retention maps for each of the sorbents are shown in Figures 30–33. Since the sorbents combine both ion-exchange and reversed-phase capability, the retention map curves are plotted with the summation of both retention mechanisms, combining the individual reversed-phase and ion-exchange curves.

Strong Cation-Exchange and Reversed-Phase Sorbent

Oasis MCX is the strong cation-exchange and reversed-phase sorbent, with the retention map [Figure 30] shown for a "weak" base compound.

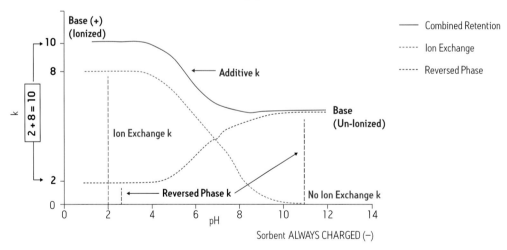

Figure 30: Oasis MCX Retention Map for a Weak Base

The reversed-phase retention mechanism is shown as a blue dotted curve, and is exactly the same as was plotted in the pure reversed-phase retention map we saw earlier in Figure 26. There is an increase in retention as the base becomes un-ionized, as would be expected.

The red dotted curve is the pure cation-exchange retention map. Since this is a strong ion exchanger, the sorbent is always charged. However, ion exchange can only occur when the compound has the opposite charge, which for a weak base is at low pH. The red curve has high retention at low pH and no ion-exchange retention at high pH because the compound becomes un-ionized.

The purple summation curve is shown for both retention mechanisms, which is what would happen to a weak base compound on this sorbent. For example, to use the capture strategy for a weak base, the sample should be loaded, and then washed at low pH to maintain the strongest retention, and then eluted at high pH. This high pH solvent will typically have a high organic concentration to be strong enough to release the compound which is now retained only by reversed-phase for an un-ionized base.

Strong Anion-Exchange and Reversed-Phase Sorbent

Oasis MAX is the strong anion-exchange and reversed-phase sorbent. The retention map, shown in Figure 31, is for a weak acid compound.

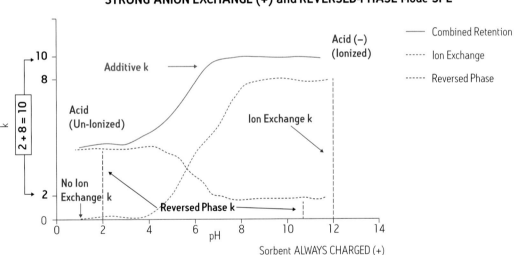

Figure 31: Oasis MAX Retention Map for Weak Acid

Again, note that both retention map curves are plotted with the dotted lines, and the summation retention map curve is shown in green for a weak acid compound. Maximum retention occurs at high pH when the anion-exchange mechanism is working. At low pH, the acid is un-ionized, so there is no ion-exchange mechanism. A capture strategy protocol for this compound would have the load and wash steps performed at high pH and the elution step at low pH.

Weak Cation-Exchange and Reversed-Phase Sorbent

Oasis WCX is the weak cation-exchange and reversed-phase sorbent. The retention map shown is for a strong base compound. [Figure 32]

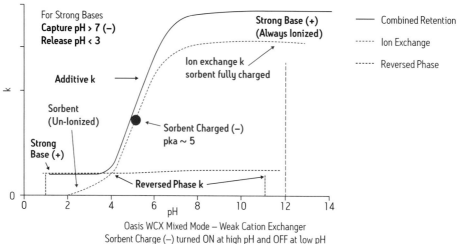

Figure 32: Oasis WCX Retention Map for Strong Base

The summation retention curve is shown in blue for a strong base compound. At low pH, the sorbent is un-ionized, which shuts off the ion-exchange mechanism. A capture strategy protocol for this compound would have the load and wash steps performed at high pH, and the compound released in an elution step at low pH.

Weak Anion-Exchange and Reversed-Phase Sorbent

Oasis WAX is a weak anion-exchange and reversed-phase sorbent, with the retention map shown for a strong acid. [Figure 33]

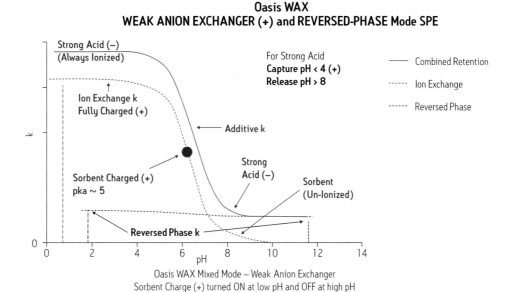

Figure 33: Oasis WAX Retention Map for Strong Acid

The summation retention curve is shown in red for a strong acid compound. At high pH, the sorbent is un-ionized which shuts off the ion-exchange mechanism. A capture strategy protocol for this compound would have the load and wash steps performed at low pH, where the maximum retention is obtained through the anion-exchange mechanism. Elution would be at high pH.

Note: Never use a strong exchanger with a strong compound because they are always both charged, and, therefore, the ion-exchange mechanism cannot be shut off. The compound would never be able to be released from the SPE cartridge. However, if a particular interfering substance is "strong", then the strong exchanger cartridge could be used to capture it completely, and then dispose of it, still captured in the SPE cartridge!

A Powerful Strategy to Simplify SPE Method Development Using Four Ion-Exchange Sorbents

Application development scientists at Waters have created a simplified method development approach, which is built upon the ion-exchange information and retention maps we have just learned. It is called the "Oasis 2 x 4 Method Development Strategy." It combines the four Oasis ion-exchange sorbents with two method protocols that provide you with a proven, logical set of experiments that result in a much faster solution to your SPE problem. Closer inspection of the four retention maps for the sorbents shows very similar retention behavior for the Oasis MCX and WAX materials, and the Oasis MAX and WCX materials. Therefore, we can combine these pairs with their own optimal protocol. Figure 34 diagrams the approach.

[Oasis 2x4 Method]

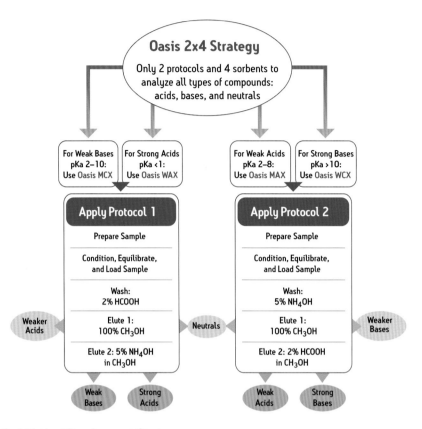

Figure 34: Oasis 2x4 Method Development Strategy

Simply determine the type of analyte[s] you are trying to capture: neutral, strong or weak, acid or base. The 2x4 strategy tells you which sorbent to select and provides a SPE protocol for you to follow.

For example, if you have a weak base, Figures 34, 35, and 36, show that the strong cation-exchange sorbent is used and you follow Protocol 1.

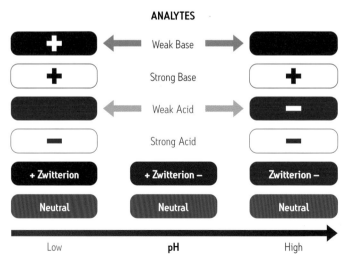

Figure 35: Charged States of Analytes Relative to pH

To capture a weak base, the sorbent bed remains ionized at both low and high pH. The weak base is captured at low pH when it is positively charged, and eluted at high pH when in its un-ionized form, thus shutting off the ion-exchange mechanism. Protocol 1 is specified.

For a clearer picture of what is actually occurring, Figures 36–40 have been prepared to aid your understanding of the science behind this powerful methods development approach.

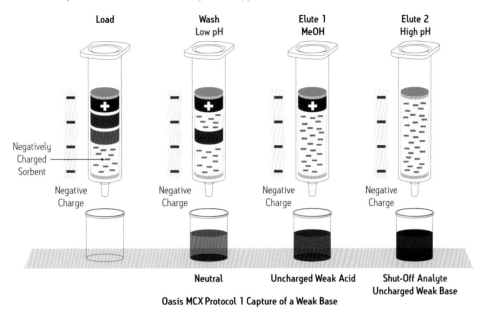

Figure 36: Capture of a Weak Base on Oasis MCX

If the goal is to capture a strong acid, use the Oasis WAX sorbent with Protocol 1. See Figures 34, 35, and 37.

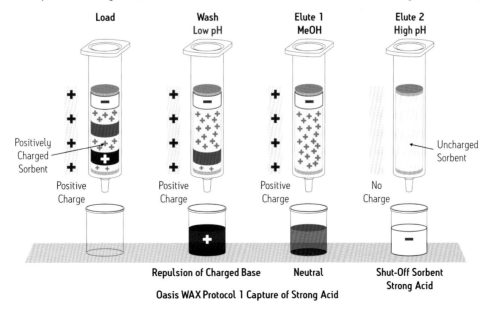

Figure 37: Capture of a Strong Acid on Oasis WAX

If the goal is to capture a weak acid, use Oasis MAX with Protocol 2. See Figures 34, 35, and 38.

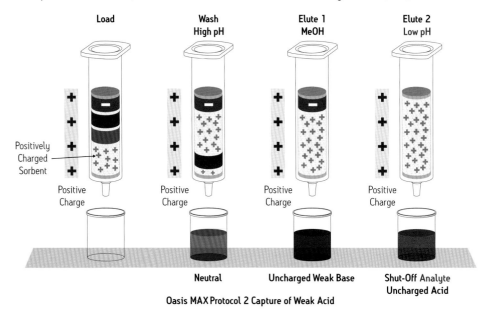

Figure 38: Capture of a Weak Acid on Oasis MAX

If the goal is to capture strong base, use Oasis WCX with Protocol 2. See Figures 34, 35, and 39.

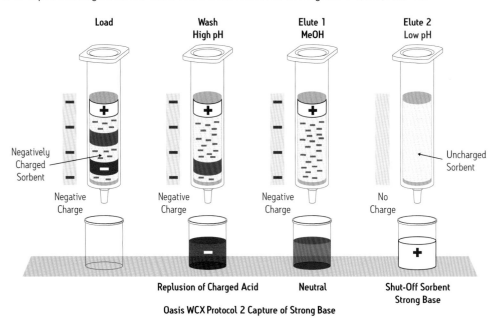

Figure 39: Capture of a Strong Base on Oasis WCX

Finally, if the goal is to capture a zwitterionic analyte, you could use Oasis MCX and Protocol 1. See Figure 40. Here we will use the positive charge on the analyte for capture, and then release with the repulsion from the two negative charges.

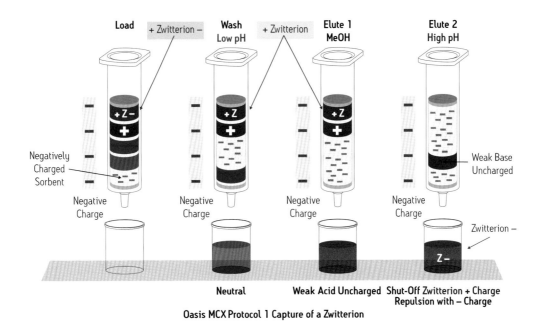

Figure 40: Capture of a Zwitterion on Oasis MCX

If you are not sure about the type of analyte you have, or if you are looking at different sample matrices, then you can still use the 2x4 strategy by creating the following scouting experiment. Run your sample on each of the four sorbents with their specific protocols and evaluate the results. Interestingly, the 2x4 strategy is very effective for Zwitterionic compounds. It will provide the best approach for a successful method. Typically one of the combinations will provide a great starting place to begin your optimization effort.

Specialty Techniques

A unique capability of SPE is to use the sorbent bed to chemically react with the sample as the sample is passed through the cartridge. This can be seen in the analysis of air for the presence of trace quantities of aldehyde- and ketone-based contaminants. [This is one area where SPE can be used on a sample that is in the gas phase.] The goal of this technology is to react specifically with those contaminants, which are not UV absorbing molecules, with a reagent [such as DNPH] that is coated onto the high surface area sorbent bed. The reagent reacts with the contaminants to form their hydrozone derivatives, which are UV absorbing molecules. These derivatives are captured by the sorbent bed and then eluted and subsequently analyzed by LC with UV detection. The following section provides more details on this process.

Sep-Pak DNPH-Silica Cartridges for Analyzing Formaldehyde Other Aldehydes and Ketones in Air

Formaldehyde and other aldehydes are receiving increased attention both as toxic substances and as promoters in the photochemical formation of ozone in air. Sources of aldehydes in residential buildings include: plywood and particle board, insulation, combustion appliances, tobacco smoke, and various consumer products. Aldehydes are also released into the atmosphere in the exhaust of motor vehicles and other equipment in which hydrocarbon fuels are incompletely

[DNPH Cartridges]

burned. The most sensitive and specific method for analyzing aldehydes and ketones is based on their reaction with 2,4-dinitrophenylhydrazine [DNPH] and subsequent analysis of the hydrazone derivatives by LC. The hydrazones may be detected by absorbance in the ultraviolet region, with maximum sensitivity obtained between 350 and 380 nm.

Airborne aldehydes have traditionally been collected by drawing a sample through an impinger containing a solution of DNPH. However, the impinger collector is generally cumbersome to use, not readily portable, and is not well suited for high flow rates or extended collection times due to solvent evaporation. Sep-Pak DNPH-silica cartridges [Figure 41] meet the requirements of United States Environmental Protection Agency [EPA] Method TO-11A and provide a convenient device for sample collection. Using a vacuum pump, an air sample is drawn through the Sep-Pak DNPH-silica cartridge. The aldehydes and ketones react with the DNPH and form the hydrazone derivative, which is retained on the cartridge. Later, the hydrazones are eluted from the cartridge with acetonitrile and analyzed by LC. Detection limits can be as low as 3 parts-per-billion volume [ppbv] for a 100 liter sample.

These cartridges provide significant advantages when compared to other techniques, such as liquid impingers, for the analysis of aldehydes and ketones. In addition, a new high speed, high resolution LC application has been developed to provide excellent quantitation capability in the low parts-per-billion range.

Figure 41: Sep-Pak DNPH Cartridges—Short and Long Body

Some of the results from sampling using these cartridges and the final analytical chromatography are shown in Figures 42–44.

HPLC Conditions

Column:	Nova-Pak® C_{18} 3.9 x 150 mm	Gradient:	100% A for 1 minute, then linear gradient from 100% A to 100% B in 10 minutes
Sample:	Mixture of DNPH and DNPH derivatives in acetonitrile.	Flow Rate:	1.5 mL/min
Mobile Phase:	A: Water/Acetonitrile/Tetrahydrofuran 60/30/10 v/v/v	Injection:	20 μL
	B: Water/Acetonitrile 40/60 v/v 3 mL	Detection:	360 nm

1. DNPH
2. Formaldehyde-DNPH
3. Acetaldehyde-DNPH
4. Acetone-DNPH
5. Acrolein-DNPH
6. Propionaldehyde-DNPH
7. Crotonaldehyde-DNPH
8. Butanone-DNPH
9. Butyraldehyde-DNPH
10. Benzaldehyde-DNPH
11. Isovaleraldehyde-DNPH
12. Valeraldehyde-DNPH
13. o-Tolualdehyde-DNPH
14. m-Tolualdehyde-DNPH
15. p-Tolualdehyde-DNPH
16. Hexaldehyde-DNPH
17. 2,5-Dimethylbenzaldehyde-DNPH

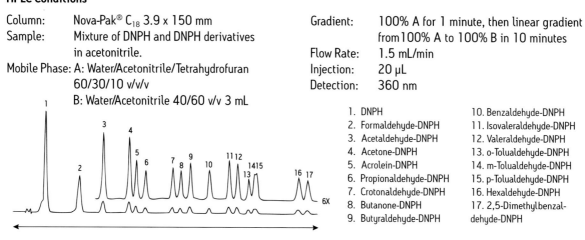

Figure 42: HPLC Separation of DNPH Derivatives of Common Aldehydes and Ketones [DNPH]

[Ozone Scrubber Cartridge]

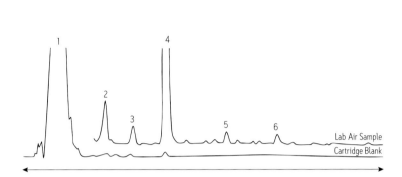

Lab Air Sample
1. DNPH
2. Formaldehyde-DNPH   4.8 ppbv
3. Acetaldehyde-DNPH   1.2 ppbv
4. Acetone-DNPH   118.0 ppbv
5. Butanone-DNPH   0.8 ppbv
6. Isovaleraldehyde-DNPH   0.7 ppbv

Cartridge Blank
[calculated for 100 liter sample]
1. DNPH
2. Formaldehyde-DNPH   0.35 ppbv
3. Acetaldehyde-DNPH   0.27 ppbv
4. Acetone-DNPH   0.34 ppbv
5. Butanone-DNPH   0.8 ppbv
6. Isovaleraldehyde-DNPH   0.7 ppbv

Figure 43: Low-Level: Aldehyde Profile from Laboratory Air

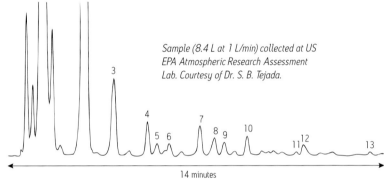

1. DNPH
2. Formaldehyde-DNPH   2290 ppbv
3. Acetaldehyde-DNPH   367 ppbv
4. Acetone-DNPH   26 ppbv
5. Acrolein-DNPH   8 ppbv
6. Propionaldehyde-DNPH   12 ppbv
7. Crotonaldehyde-DNPH   36 ppbv
8. 2-Butanone-DNPH   24 ppbv
9. Butyraldehyde-DNPH   15 ppbv
10. Benzaldehyde-DNPH   32 ppbv
11. o-Tolualdehyde-DNPH   5 ppbv
12. m-p-Tolualdehyde-DNPH   11 ppbv
13. 2,5-Dimethylbenzaldehyde-DNPH 3 ppbv

Sample (8.4 L at 1 L/min) collected at US EPA Atmospheric Research Assessment Lab. Courtesy of Dr. S. B. Tejada.

Figure 44: High-Level: Aldehyde Profile from Diluted Auto Exhaust Emissions

Ozone Scrubber Cartridges

Ozone has been shown to interfere with the analysis of carbonyl compounds in air samples that have been drawn through cartridges containing silica-coated with DNPH. Waters Ozone Scrubber cartridges are designed to remove this ozone interference. These disposable devices are intended for use in series combination with the Waters Sep-Pak DNPH-silica cartridges or XPoSure™ Aldehyde Sampler cartridges, on page 58. One ozone scrubber cartridge replaces the one 4" diameter by 36" long copper ozone denuder located in the heated zone of sampling systems used for outdoor air monitoring [PAMS program]. Each ozone scrubber cartridge contains 1.4 g of granular potassium iodide. When air containing ozone is drawn through this packed bed, iodide is oxidized to iodine, consuming the ozone. The theoretical capacity of a single cartridge is 4.2 mmoles of ozone [200 mg]. The particle size of the potassium iodide granules is optimized for good mass transfer and flow characteristics. See Figures 45 and 46.

[XPoSure Cartridge]

Figure 45: Ozone Scrubber Cartridge

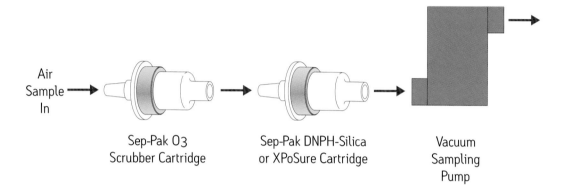

Figure 46: Flow Schematic for Air Sampling System

Waters XPoSure Aldehyde Sampler Cartridges for Monitoring Aldehydes in Indoor Air with Personal Portable Pumps

Based on an extension of our DNPH coating technology, XPoSure aldehyde sampler cartridges are designed to work with portable personal sampling pumps. They feature very low back pressure and very clean background levels to provide high sensitivity results.

Using this cartridge is as easy as sample, elute, and shoot. There's no need to break open and manipulate glass tubes. Because the cartridges are made from high-density polyethylene (HDPE), breakage is not a concern. Figures 47 and 48 show the cartridge and an analyte trace. An actual cartridge blank demonstrating extremely low background levels and an actual laboratory air sample are shown. Also, note that in one sample, several different compounds can be analyzed at the same time.

Figure 47: Waters XPoSure Aldehyde Sampler

[PoraPak RDX]

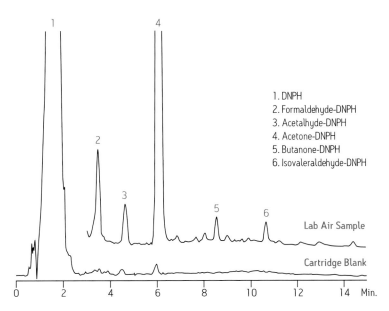

Figure 48: Low-Level Example: Aldehyde Profile from Laboratory Air Using XPoSure Aldehyde Samplers

PoraPak Rdx Sep-Pak Extraction Cartridge for the Analysis of Explosives in Surface and Ground Waters

Designed to meet or exceed the QA/QC requirements of EPA Method 8330, this cartridge is used for environmental testing laboratories involved with Department of Defense [DoD] remediation programs.

PoraPak® Sep-Pak cartridges contain PoraPak Rdx resin, a specially prepared, specially cleaned divinylbenzene/vinyl-pyrrolidone copolymer, packed in a high purity polypropylene syringe barrel. With the lowest guaranteed backgrounds and the highest cartridge-to-cartridge, lot-to-lot consistency. The PoraPak Rdx cartridge is the most sensitive technology available today and allows you to perform analysis at sub ppb levels. This resin is highly sensitive for nitroaromatic and nitroamine compounds resulting in recoveries of 90% or greater. [Figures 49 and 50]

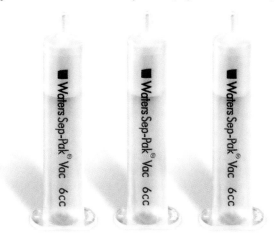

Figure 49: PoraPak Rdx Cartridges

[Sep-Pak Dry Cartridge]

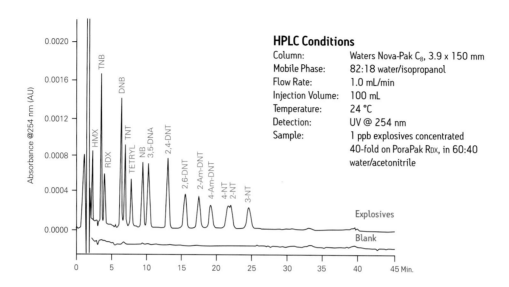

Figure 50: Isocratic Separation of Method 8330 Analytes [PoraPak Rdx]

Sep-Pak Dry SPE Cartridge

Sep-Pak Dry cartridges are packed with 2.85 g of anhydrous sodium sulfate. These cartridges are designed to remove residual water from the SPE extract. See Figure 51.

Figure 51: Sep-Pak Dry Cartridge

Key Terms
and Calculations

DETERMINING THE LIQUID HOLD-UP VOLUME OF A CARTRIDGE

Every solid-phase extraction [SPE] device has what is called the hold-up volume, or the volume of liquid that remains inside the device when the flow is stopped. Knowing this volume of liquid will be important to ensure success during the method development process. See Figure 52.

Understanding the Concept

The hold-up volume exists within the pores of the packed bed of sorbent particles, the spaces between the particles, the pores of the two filters, and the inside volume of the outlet fitting [tip]. For a typical SPE cartridge, a significant portion of this hold-up volume is created by the open spaces between each sorbent particle. This space between the sorbent particles is called the interstitial space or the interstitial volume in the sorbent bed.

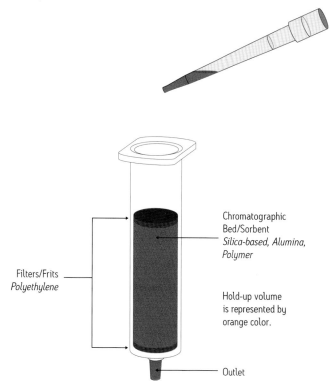

Figure 52: SPE Device Hold-Up Volume

To help understand this concept, think about a large drinking glass filled with crushed ice. The drinking glass itself represents the body of the SPE cartridge, and the crushed ice represents the sorbent particles. When the glass is then filled with water, the water fills the interstitial spaces between the crushed ice particles. When you quickly drink this glass of "ice water," you are drinking the interstitial volume of water that was held in that drinking glass between all the crushed ice particles. See Figure 53.

[Determining the Hold-Up Volume]

Figure 53: Familiar Example of Interstitial Volume

Therefore, the total hold-up volume for an SPE cartridge will depend upon several factors: the dimensions of the packed bed and particle size of the sorbent, its interstitial volume, the liquid volume in the pores of the sorbent particles and the two filters, and finally the volume of the outlet male fitting. The orange shaded portion of the cartridge in Figure 52, represents the hold-up volume.

Determining the Hold-Up Volume

SPE cartridge manufacturers are sometimes able to provide the hold-up volume information. For more information, refer to SPE cartridge vendors websites. However, you can also determine it empirically by setting up a cartridge using a solvent containing a dye for easier visual inspection. Note: Do not use water for this determination because with certain sorbents [i.e. C_{18}], it will not wet all the pores of the sorbent particles. This will give an inaccurate hold-up volume. Refer to the "Conditioning and Equilibration" steps discussed in the "SPE is LC" section, on page 35.

Slowly pass the dye solution into the SPE device, and stop the flow just as the first drop is about to exit the outlet tip. Measure exactly how much volume was loaded onto the cartridge. Note, if an open syringe style cartridge is used, and there is residual dye solution above the sorbent bed and top filter, decant and measure the residual volume above the sorbent bed, then subtract that value from the volume determined in the load step. The net volume is the approximate overall liquid hold-up volume of the cartridge.

The hold-up volume is important when trying to determine whether all, or some portion of, the compound of interest is still inside the SPE device. For example, you are working with a sample that contains a compound of interest which has a k = 0 [no retention], under the chromatographic conditions in the loading step. See the Glossary for more information on the Retention Factor k, on page 187. If 1 mL of this sample is loaded onto a SPE cartridge with a hold-up volume of 2 mL, then all of the compound of interest must still be inside the cartridge. Remember, it would take more than 2 mL of this sample load before the sample compounds could reach the outlet of the cartridge.

However, if a smaller SPE cartridge, with a hold-up volume of only 0.5 mL, is used and 1 mL of sample is still loaded onto the cartridge, then half of the compound of interest with a k = 0, has already exited from the cartridge during the loading step, and the other half is still inside the cartridge. It will be important to keep this concept in mind in order to develop robust SPE methods.

In some applications, it may be useful to remove the interstitial liquid between wash and elution steps [after the load step has been completed]. Use vacuum to draw in air, or use centrifugation to expel all of the interstitial liquid. However, remember that some liquid will still be present in the sorbent pores. We can use vacuum to draw in air at this point, since the sample has already been loaded and captured by the sorbent. Drawing air into the cartridge must be avoided before the sample is loaded.

Flow Rate and Linear Velocity

In the "SPE is LC" section, we learned that SPE technology is essentially liquid chromatography used for sample preparation. The fundamental principles of the two technologies are the same. In both approaches, the different compounds in a sample matrix can be separated by a chromatographic bed of particles and a flowing stream of solvent or mobile phase.

The separation power of the chromatographic bed depends on several factors, including chemistry of the sorbent particles [i.e. C_{18} or CN], particle size, length of the bed, type of solvent [strong or weak], resultant k value of the analyte, and speed of the mobile phase or linear velocity. It is important to understand the effects that the speed of the mobile have on SPE performance.

Understanding Linear Velocity

The speed of the flowing solvent is called its linear velocity, and this value is important in determining the separation power in liquid chromatography. In general, if the linear velocity is too high, the separation performance will be decreased, sometimes dramatically. Slower linear velocities than specified in the SPE method do not adversely impact the performance of the separation; they just result in more time to process the samples.

Typically, linear velocity is not specified in a method; however, the flow rate and cartridge size are specified, and together these two values define the proper flow velocity for the method. It is important to know the volume of liquid for each step and the time it should take to be applied to the SPE cartridge to ensure proper performance.

For example, a well developed, effective SPE protocol will stop working if the liquids are allowed to flow faster than specified during a protocol run.

In a SPE method, there are specific steps that require flow speed control. See Table 5.

Table 5: Steps Requiring Flow Rate Control

Step	Liquid
Load	Sample/sample solvent
Wash	Wash solvent
Elute	Elution solvent

Note that the conditioning and equilibration steps do not require flow control. See Figure 54.

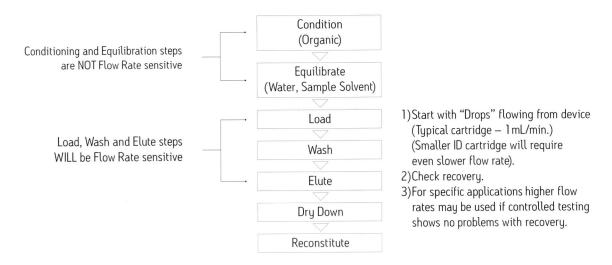

Figure 54: Load, Wash, and Elution Steps Require Flow Rate Control

However, in SPE methods, we don't think about the linear velocity [distance/time] of the flowing liquids; instead, we think about their flow rates [volume/time]. The two values are related, and are a function of the cross sectional area [internal diameter] of the packed chromatographic beds of sorbent/particles.

For those familiar with LC, a separation method will specify the flow rate of the mobile phase, which is set by the LC pump. The internal diameter [ID] of the LC column will actually determine the specific linear velocity of the mobile phase for that flow rate.

If the flow rate of mobile phase is increased by three times, then the linear velocity will go up three times. In LC, a three times increase in linear velocity will usually result in very poor separations on the resulting chromatogram.

Importance of Flow Rate Control

Consider the loading step for a green sample and the impact of flow control. The protocol in this example specified that 2 mL of sample matrix volume were to be loaded onto the sorbent bed in a SPE cartridge. We can calculate the flow rate of the load step by measuring how much time it takes to load that sample volume. See Figure 55.

For example, for sample one [on the left], it takes two minutes to load 2 mL, which means the flow rate is 1 mL/min. However, a second sample [on the right], is loaded in only 10 seconds, so the flow rate is now 12 mL/min., which is a 12 times increase in flow rate. Since the SPE cartridge size is the same for both samples, the effective linear velocity

[Scaling and Linear Velocity]

for the second sample was twelve times higher! The separation performance on the second sample would be very poor, due to the significant increase in the linear velocity during the load step. As mentioned earlier, the wash and elution steps are also impacted by changes in the linear velocity.

General guideline: You should observe "drops" exiting from the SPE cartridge. This is a good indicator that the linear velocity [flow rate] is acceptable [see left diagram on Figure 55]. If a "steady stream" of liquid is seen exiting the cartridge during the loading step [see right diagram on Figure 55], then that linear velocity is too high. This will result in poor capture of the compounds of interest, causing a premature "breakthrough" and poor recovery.

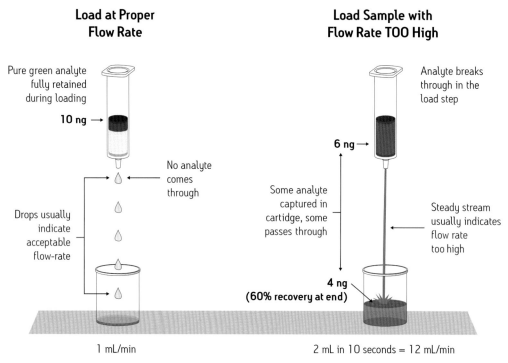

Figure 55: Importance of Flow Rate Control—Load Steps

Scaling and Linear Velocity

It is also important to understand that a change will occur in the linear velocity of the mobile phase when the cartridge ID is changed, even if the flow rate remains the same! The ratio of the cross sectional areas of the two different sorbent/particle bed dimensions will determine the actual linear velocity of the liquid in the scaled SPE cartridge. Table 6 provides the general relationship between cartridge ID and linear velocity.

[Scaling and Linear Velocity]

Table 6: Relationship Between Cartridge ID and Linear Velocity*

At the Same Mobile Phase Flow Rate	
Cartridge/Column Internal Diameter [Original to Scaled]	Impact on Solvent Linear Velocity [Original to Scaled]
Smaller ID to Larger ID	Goes DOWN
Larger ID to Smaller ID	Goes UP

Flow rate held constant

Here is an example of scaling and how we adjust for linear velocity and how we adjust for sample and solvent volumes. See Figure 56.

Existing application: Developed on a 3 cc with 200 mg
1.0 mL sample volume loaded in 1 minute at 1.0 mL/ min.

Scaled application: A larger 20 cc with 5 g cartridge

The first thing we calculate is the "scaling factor":.

$$SF = \frac{\text{New sorbent mass}}{\text{Original sorbent mass}} = \frac{5 \text{ g}}{200 \text{ mg}} = 25x$$

Next, the scaled sample volume = SF x original sample volume = 25 x 1.0 mL = 25 mL

We can see that the sample volume will be 25 mL, but how much time will be required to keep the linear velocity the same and what flow rate is required?

Figure 56: Scaling from Smaller to Larger SPE Device

Since the new cartridge sorbent bed cross sectional area, [πr^2] is ~5 times larger, the linear velocity will go down 5x, if the flow rate is kept at 1.0 mL/min. To maintain the same linear velocity, higher loading flow rates can be investigated, potentially up to 5 mL/min. Therefore, 25 mL of sample in 5 minutes would be appropriate. Always check separation performance by measuring % recovery values. [See Calculating Recovery on page 69.]

[Scaling and Linear Velocity]

However, if the goal of the scaling process is to go down in sample volume and cartridge size, especially with smaller sorbent bed cross sectional areas, then the flow rate will have to be decreased to maintain the proper linear velocity. See Figure 57, for example.

> Existing application: Developed on a 3 cc with 500 mg cartridge
>
> 1.0 mL sample volume loaded in 1 minute at 1.0 mL/min.
>
> Scaled application: Smaller Sep-Pak Light cartridge with 120 mg sorbent
>
> [Scaling factor = 120mg/500mg = 0.24]

The sample volume will be 0.24 mL, but in how much time? What is the flow rate that will be required to keep the linear velocity the same?

3 cc   Light

Figure 57: Scaling from Large to Smaller Device

Since the new cartridge has a much narrower sorbent bed [~1/3], with a ~ 1/7 smaller cross-sectional area, the loading flow rate must be decreased by 1/7th. In this case, investigate lower flow rates from 0.1 to 0.2 mL/min. Therefore,

$$\text{Load Time} = 0.24 \text{ mL} \div 0.15 \text{ mL/min} = \sim 1.6 \text{ minutes}$$

As above, always check for proper % recovery performance.

Note: Some modes of chromatography, especially those using ion-exchange sorbents, require even lower linear velocities than normally seen in other applications. SPE methods using ion-exchange chromatography will require lower linear velocities [lower flow rates] for proper separation performance.

A part of the robustness assessment of a SPE protocol will be the determination of what linear velocities will be acceptable. The protocol should clearly state how to maintain control of the liquid speeds by referencing the related flow rates that are allowed. Clearly specify the volume of liquid, and the time required to apply that volume.

Calculating Recovery

The most commonly used term to evaluate how well a SPE method performs is called the "% recovery" of the method. In its simplest form, recovery is the ratio of mass released [how much mass of that compound of interest was released from the SPE cartridge], to mass loaded [how much mass was originally loaded onto the cartridge]. This is given as a percentage rating from 0 to 100%, where 100% recovery is the ideal result, indicating that all the mass that was loaded was then released.

For example, Figure 58 diagrams a simple method where 10 ng of a pure "green" compound of interest was loaded onto the cartridge. A wash step was performed properly so that none of the compound was released. The elution step was performed, and 10 ng of the "green" compound was recovered from the cartridge.

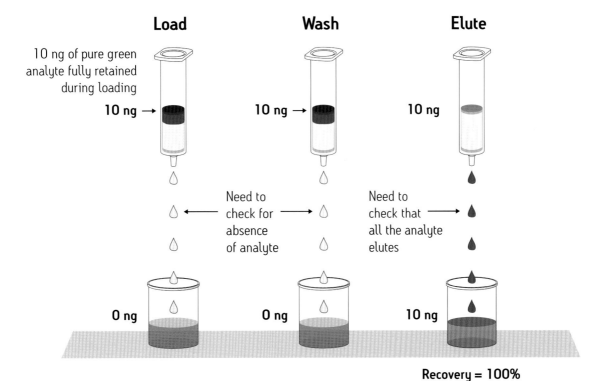

Figure 58: Determining % Recovery

The simple recovery for this method is calculated as follows:

$$\text{Recovery} = \frac{\text{mass recovered}}{\text{mass loaded}} \times 100$$

$$\%\text{ Recovery} = \frac{10 \text{ ng}}{10 \text{ ng}} \times 100 = 100\%$$

[Improved % Recovery Calculation]

A more scientifically correct way to calculate the % recovery is performed in two parts using the actual sample matrix and pure standards for the compound[s] of interest. The benefit of this calculation is that the detected amount of the compound eluted from the SPE cartridge may be impacted by problems in the load, wash, and elution steps, or by any problems from the SPE device, or variation in execution of the method.

Adjusting for Errors from the SPE Device or Protocol

The first step is to create a control sample, as shown in the left diagram of Figure 59. The sample matrix without the compound[s] of interest, which is called the blank sample matrix, is processed through the SPE cartridge, just as if it were the actual sample. The load, wash, and elution steps are performed normally. During the elution step, what exits the SPE cartridge will be collected and labeled "post-extracted sample." This post-extracted sample vessel will contain any substances still present from the blank sample matrix. To this vessel, add the standards for the compounds of interest at the appropriate concentration [here 10 ng/mL]. Adjust the labeling to read, "post-extracted spiked sample." Analyze this post-extracted spiked sample for the compounds of interest. Since the compounds of interest did not go through the SPE protocol, we now have an accurate reported value for them in the SPE processed blank sample matrix. Here we added enough standard to make 10 ng/mL in the sample vial, and the detected response was 10 ng/mL.

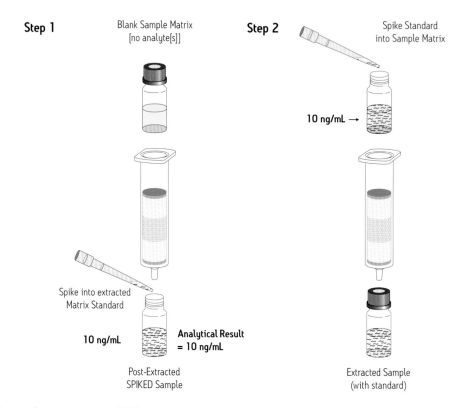

Figure 59: Proper Determination of % Recovery

The second step is to perform the SPE protocol with the compounds of interest in the sample matrix from the beginning. Refer to the diagram on the right of Figure 59. Note: The mass of the standards added to the first and second experiments must result in the same concentration for both.

First, add the standard of the compound of interest to the blank sample matrix, and label the vessel as "spiked sample matrix." Now perform the same SPE protocol. During the elution step, collect what exits the cartridge and label it "extracted sample with standards," and analyze as before. This will allow us to determine if any of the compounds of interest are not fully released during the SPE protocol.

The calculation is shown below in Figure 60. The % recovery can then be determined for each of the compounds of interest that were studied.

Here we obtained a detected value of 9.5 ng/mL for our standard in the extracted sample, while the post-extracted spiked sample was 10 ng/mL. Using the equation, this results in a 95% recovery.

$$\% \text{ RE}^* = 100 \times \left(\frac{\text{Response Extracted Sample (with standard)}}{\text{Response Post-Extracted SPIKED Sample}} \right)$$

*Recovery of the Extraction Procedure RE

$$\% \text{ RE}^* = 100 \times \left(\frac{9.5}{10.0} \right) = \textbf{95\% Recovery}$$

Figure 60: Calculating % Recovery

This recovery calculation gives the most reliable value and ensures that the SPE protocol is working and is being evaluated properly for the true recovery of the compounds of interest.

In general, a 100% recovery rating for a method is ideal, and especially in the pharmaceutical industry where a goal of >95% recovery is typically desired. However, in other areas, such as environmental samples, >75% recovery may be satisfactory. Percentage recovery guidelines will be available to you as methods development work begins. When performing an SPE method on a routine basis, the % recovery is used to indicate if the method is working properly.

Note: In some published methods, reported % recoveries of >100% are seen. This does not mean that the SPE protocol is creating more mass for that compound. It typically is the result of minor measurement errors in the calculation of the mass loaded and/or in the determination of the mass recovered values.

Also, if using MS detection this calculation will not show any matrix effects from the sample. If any suppression or enhancement is present from the sample/ matrix, it will be normalized. To detect matrix effects, the following experiment is performed.

Matrix Effects Calculations

The next important calculation determines whether the sample matrix itself is causing a problem in the accurate determination of the mass of the analyte when using a mass spectrometer. This is called the sample "matrix effect." In LC-MS, the MS detection response for the compound of interest can be altered by the presence of substances from the sample matrix. If the response is reduced, it is called ion suppression, or if increased, it is called ion enhancement. In either case, the reported mass for the compound will be incorrect. This is important because the response for the compound is used to determine how much mass of that compound was present. If the response is not accurate, then the reported mass value for that compound will not be accurate. This will result in an inaccurate % recovery calculation for the SPE method, since the altered detected response of the compound will be incorrectly attributed to poor performance of the SPE protocol.

The following procedure will identify any response errors and calculate the degree of ion suppression or ion enhancement effects. These effects may be the cause of poor % recovery results, even though the SPE method was performing well for non-MS detection system.

A properly designed experiment can clearly identify whether the sample matrix is causing any reporting errors in the amount of analyte present. In some cases, there is no accuracy problem. To determine any matrix effects, the experiment and calculation shown in Figures 61–64 should be performed, during method development.

As shown in Figure 61, two samples are prepared and analyzed. First, as shown on the left, a blank sample matrix [no analytes present] is processed through the SPE device according to the desired method, and then the analyte[s] are spiked into the vial. Note: this is the same first step used in the % recovery calculation. On the right, the second sample is prepared by taking the pure analyte standard[s] and spiking them into the same, pure solvent [neat solvent, no sample matrix] that was used in the elution step to obtain the post-extracted spiked sample. It is important that the neat solvent for step 2 is the same solvent used as the elution step for the first experiment. Adjust the amount of the spiked standard to give the same concentration of standard as in Step 1.

[Matrix Effects Calculations]

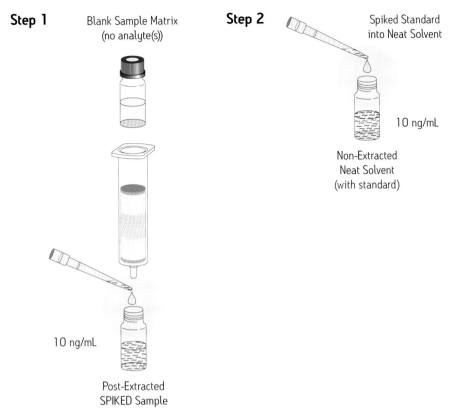

Figure 61: Proper Determination for Matrix Effects

The two analyte response values from the mass spectrometer are then put into the formula shown in Figure 62. This will provide a numerical assessment of the effect that the remaining sample matrix interferences are having on the response. Since the post-extracted spiked sample had been processed through the SPE device before the standards were added, any response errors are due to whatever sample matrix components were still present. Three scenarios can result, as shown in Table 7.

Table 7: Matrix Effects Scenarios

Response of Post-Extracted Spiked	Response of Non-Extracted Spiked	Matrix Effect
Same	Same	0%
Lower	Higher	% Ion Suppression
Higher	Lower	% Ion Enhancement

[Matrix Effects Calculations]

$$\% \text{ ME}^* = 100 \times \left(\frac{\text{Response Post Extracted Spiked Sample (with standard)}}{\text{Response NON-Extracted Neat Solvent with analyte(s)}} - 1 \right)$$

* Matrix Effects (ME)
- Both samples should be in the same solution
- Negative value = Suppression
- Positive value = Enhancement

$$\% \text{ ME} = 100 \times \left(\frac{10.0}{10.0} - 1 \right) = 0\%$$

**If both results are the SAME,
% ME = 0%
No matrix effect**

Post-Extracted Spiked	Non-Extracted Neat	Matrix Effect
10 ng/mL	10 ng/mL	0%

Figure 62: First Scenario: No Matrix Effects

When both response values are the same, the calculation results in a 0% matrix effect.

If the post-extracted spiked value after SPE is lower than the neat solvent value, then the residual sample matrix interferences after the SPE protocol are causing ion suppression. As shown in Figure 63, for this experiment there is a significant level [80%] of ion suppression.

[Matrix Effects Calculations]

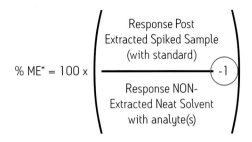

* Matrix Effects (ME)
- Both samples should be in the same solution
- Negative value = Suppression
- Positive value = Enhancement

$$\% \ ME = 100 \times \left(\frac{2.0}{10.0} - 1 \right) = -80\%$$

% ME = -80%
which means 80% Suppression

Post-Extracted Spiked	Non-Extracted Neat	Matrix Effect
2 ng/mL	10 ng/mL	80% Ion Suppression

Figure 63: Second Scenario: Significant Ion Suppression

The negative % value indicates ion suppression.

[Matrix Effects Calcucation]

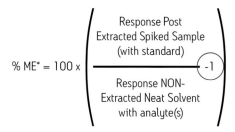

$$\% \text{ ME}^* = 100 \times \left(\frac{\text{Response Post Extracted Spiked Sample (with standard)}}{\text{Response NON-Extracted Neat Solvent with analyte(s)}} - 1 \right)$$

* Matrix Effects (ME)
- Both samples should be in the same solution
- Negative value = Suppression
- Positive value = Enhancement

$$\% \text{ ME} = 100 \times \left(\frac{12.0}{10.0} - 1 \right) = 20\%$$

% ME = 20%
which means 20% Enhancement

Figure 64: Third Scenario: Ion Enhancement

If the post-extracted spiked value after SPE is greater than the neat solvent value, then ion enhancement is present. For the experiment shown in Figure 64, a 20% enhancement effect is calculated. This means that there is some compound[s] present in the SPE processed sample matrix that is providing a greater signal from the mass spectrometer for our analyte.

Post-Extracted Spiked	Non-Extracted Neat	Matrix Effect
12 ng/mL	2 ng/mL	20% Ion Enhancement

The combination of accurate values for % recovery and matrix effects is very important when creating a high performance SPE sample preparation method.

In The Lab

[Solid-Phase Extraction Device Designs]

Solid-Phase Extraction Device Designs [Cartridge/Configurations]

If you are following an existing solid-phase extraction [SPE] method, you should use the type and size of device specified by the methods development chemist.

SPE devices come in many different designs, called "configurations," as shown in Figure 65. These can be grouped into the following types: "off-line" and "on-line" devices. This is an overview so that you can understand how your method development chemist selected the SPE device that you will be using.

Off-Line Devices

Straight sided syringe barrels and expanded reservoir Vac RC devices are [see Figure 67] used for individual or parallel sample preparation where the sample and solvents are processed through the devices typically with the help of a vacuum, and a vacuum manifold, "off-line" from the analytical instrument.

Higher throughput formats such as 96-well plate designs permit the parallel processing of many samples using vacuum or positive pressure manifolds.

Enclosed design cartridges, which have both Luer® taper inlet and outlet fittings, such as the Sep-Pak Plus, Sep-Pak Classic, and Sep-Pak Light configurations, can utilize a syringe to push the sample and the processing solvents through the cartridge, and they can also be used with a vacuum manifold.

On-Line Devices

Specially designed "on-line" steel and cartridge columns are used in the flow path of analytical instruments.

A more detailed description of the tools for using the devices can be found starting on page 85.

Figure 65: SPE Device Configurations

Off-Line and On-Line Processing

As mentioned above, there are two types of SPE processing to consider: off-line and on-line.

In off-line SPE, the samples are processed separately from the analytical instrumentation. For example, from one to several hundred samples are all prepared by the SPE protocol, before they are loaded into an analytical instrument for further analysis.

In on-line SPE, specially designed cartridges, instrument components, and columns are used within the flow path of the analytical instrument. Each sample is prepared as it is processed through the analytical instrument system.

Off-line SPE is the more popular approach, so we will begin with those configurations.

Off-Line Processing

Each configuration for off-line processing offers different construction features, to meet the varying needs of analytical scientists in sample preparation. Some designs are useful when you have only a few samples that can be processed one at a time. Some are useful if you have many samples to process simultaneously. Some are designed to process a large sample volume, such as a liter, while others are specially designed for very small sample volumes of less than 100 microliters [µL]. And finally, some designs are able to work with automated processing equipment.

Understanding the benefits of the different configurations will make your sample preparation process much more successful. The more popular configurations are described in the next section. How to use these cartridges and the different ways to process samples through them will be discussed in the "Driving Forces used to Generate Liquid Flow" section. See page 90. Refer to the Waters website for listing of all the different configurations and sizes availible.

Syringe Style

Syringe Barrel Design

The most common SPE configuration is the syringe barrel design. It has a straight sided barrel design and is usually lower in cost, for a given amount of sorbent mass. The outer syringe housing body is typically made of polypropylene; however, some specialty devices are made from glass, if plastics cannot be used with the sample. The top [inlet] will typically have flanges at the top; however, some are specially made without flanges to fit certain processing equipment. The outlet is shaped as a male Luer taper tip, which is useful when connecting the cartridge to various manifolds or other devices that have female Luer connectors. The empty internal volume for syringe style cartridges, range from 1 cc to 35 cc or mL of liquid.

Internally, a lower filter and upper filter, also called "frits", are required to contain the sorbent [stationary-phase particles]. These filters are made of polyethylene, or pure Teflon®, if a glass body is used. The nominal rating for these filters is approximately 20 µm, since most sorbents range from 50 µm and higher. See Figure 66.

[Syringe Barrel Design]

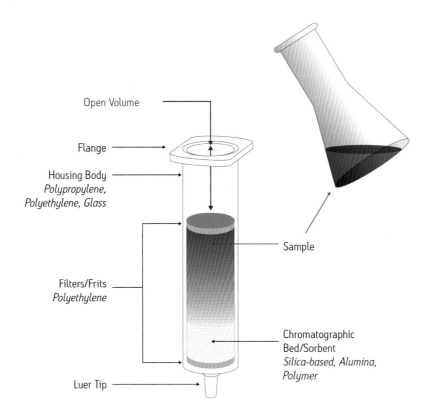

Figure 66: Diagram of Syringe Barrel Design

The sorbent is maintained between the two filters. Depending on the size of the cartridge, from 10 mg to 10 g of sorbent will be contained. The overall size required for the SPE device and its sorbent mass will be determined by the volume and complexity of the sample matrix.

Note: Syringe barrel cartridges are typically not filled to the top with sorbent. The remaining open volume can be used to temporarily hold the liquids that will be applied to the cartridge. For example, a 3 cc cartridge with 60 mg of sorbent will have approximately 2 cc of open volume.

Vac RC design

A variation of the syringe barrel design has an expanded reservoir, as shown in Figure 67, allowing larger volumes of solvents to be charged to the cartridge at one time. This is useful if the method calls for the addition of more liquid than the open volume of a straight sided syringe barrel design. The expanded reservoir variation was designed for use with early automated SPE robotic equipment. Some cartridge manufactures call these "robotic compatible" cartridges, but Waters uses the term Vac RC.

Note that the amount of sorbent and the chromatographic bed ratio of length to width can be the same in both the straight barrel syringe and the expanded reservoir syringe.

Figure 67: Expanded Reservoir Barrel Design, Vac RC

96-Well Plate Designs

Another common configuration for processing hundreds or even thousands of samples per week is based on the familiar, standard-dimension 96-well plate [8 x 12] format. See Figure 68. In this format, each "well" behaves just like an individual syringe cartridge, but with the ability to process 96 individual samples, one per well, all at the same time in a parallel process mode. Each well has the same, filter-sorbent-filter arrangement as in the syringe style configuration. These 96-well SPE plates use specially designed vacuum manifolds or automated equipment to process the liquids. Collection plates with 96 wells are used to receive the liquids during the steps of the SPE method.

Figure 68: 96-Well Plates

A patented feature of the Waters 96-well plates allows for the positioning of the two filters in different locations depending on the amount of sorbent needed for the application. For example, Figure 69 shows how different bed configurations can be achieved. This two stage well design can achieve a sorbent mass as low as 5 mg, with up to 60 mg for the high capacity, copolymer-based Oasis sorbents.

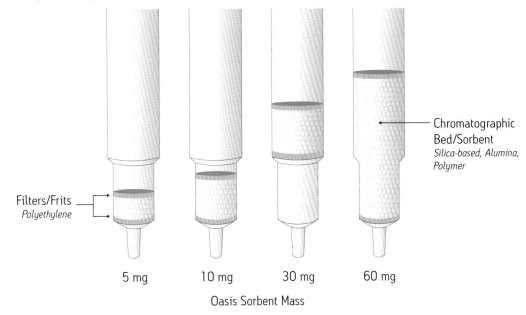

Figure 69: 96-Well Plate Bed Configurations for Oasis Sorbents

[µElution 96-Well Plate Design]

For traditional silica-based sorbents, [C_{18} silica], 25 mg up to 100 mg beds can be created, as shown in Figure 70. All 96-well plate devices are also compatible with most liquid-handling robotic systems for the automated processing of large numbers of samples.

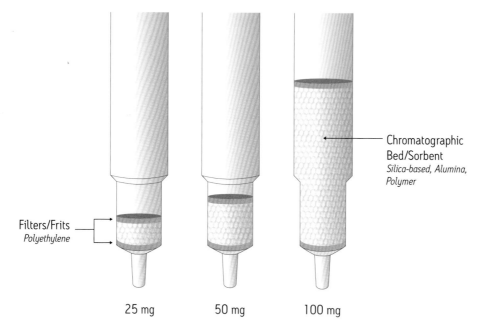

Figure 70: 96-Well Plate Bed Configuations for Silica-Based Sorbents

µElution 96-Well Plate Design

This innovative, patented configuration is specifically designed by Waters for very limited, sample volume applications [10–375 µL], where successful matrix clean-up and/or compound concentration can be achieved with final elution volumes as low as 25 µL! As seen in Figure 71, the wells are specially constructed with a sharply tapered tip outlet design.

Figure 71: Waters Oasis µElution Plate

Within this tapered tip, innovative, spherical filters are used to confine a small mass of sorbent in the filter-sorbent-filter arrangement. This can be seen in Figure 72. The amount of sorbent is specifically chosen for the small sample volumes to be processed.

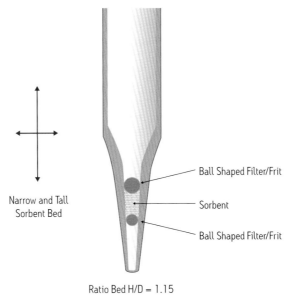

Figure 72: Patented µElution Plate Tip Design

Because of the small volume of elution solvent needed, the plate produces extracts with increased compound concentration, which can then be directly injected into the analytical instrument for faster, higher sensitivity results. The special ratio of bed height to bed diameter [H/D] creates a very efficient chromatographic bed to improve performance. This 96-well plate configuration is also compatible with most liquid-handling robotic systems for the automated processing of large numbers of samples.

Enclosed Design Cartridges

Syringe barrel and 96-well plate designs are considered to be "open" cartridges, since their inlets do not have any specific connector design. "Closed bed" cartridge designs, which have both inlet and outlet fittings have more flexibility in how they can be connected. However, you should be not exceed ~70 psi in operating pressure during use.

Classic Design

This was the first commercially available, solid phase extraction cartridge design, which was introduced by Waters Corporation in 1978. See Figure 73. This patented configuration uses a heat shrinkable polyethylene body that encloses the filter-sorbent-filter arrangement to structure and compress the chromatographic bed.

Figure 73: Classic Cartridge

[Plus and Light Designs]

As the tubing shrinks, it forms a radially-compressed, tightly-packed sorbent bed, and, both an inlet and outlet female Luer taper connector that can mate with male Luer connectors. Two different sorbent bed lengths are available depending on the type of packing material. These are called the long body and short body configurations. These cartridges are ideally suited for applications that do not require a large number of samples to be processed.

Note: the Classic cartridges are not designed for use in robotic/automated SPE equipment.

Plus Design

This configuration design features a molded polyethylene body that includes a female Luer inlet and a male Luer outlet fitting for maximum versatility and convenience. Because of the tight dimensional tolerances from the molding process, these devices are ideal for use with robotic/automated equipment. See Figure 74. The filter-sorbent-filter arrangement is enclosed by the body.

Note that the Plus configuration also has the long body and short body bed lengths in order to match the sorbent bed volumes of the Classic design, allowing them to be used interchangeably. For example, if you are using a short body C_{18} Classic Sep-Pak cartridge that contains ~360 mg of sorbent, then the short body C_{18} Plus Sep-Pak cartridge also contains ~360 mg of sorbent. A color-coded, compression ring seals the Plus body components together and indicates general sorbent types.

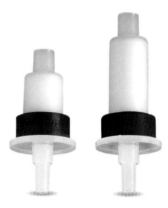

Figure 74: Plus Cartridges

Outlet tips feature reduced internal volume for minimal sample hold-up volume. In addition, female Luer inlet and male Luer outlet fittings allow for the cartridges to be stacked in tandem, with cartridges of the same sorbent for increased capacity; with cartridges of different sorbents to create sequential adsorption strategies; or with membrane filters for removal of sample particulates.

Plus design cartridges can be used with liquid and gas phase samples. They have a unique capability for reverse flow direction, a useful feature for trace enrichment applications where the trace concentrated compound, captured on the sorbent bed inlet, can be eluted in much less solvent volume by flowing the elution solvent in the reverse direction. This will be shown in a later example. See page 136.

Light Design

The Light design, Figure 75, resembles the appearance of the short body Plus configuration, except that it's outer body has a ribbed/finned surface to indicate that the internal volume/sorbent mass is much less [there is no "long body" version]. Since the Light configuration has ~1/3 the internal volume of the Plus design it is ideal, when your sample volume is limited, or when excessive dilution is a concern. Fractions can be eluted in lower volumes to improve sensitivity and reduce solvent consumption.

Figure 75: Light Cartridge

As a comparison, a short body Sep-Pak Plus cartridge contains ~360 mg of C_{18} sorbent. The Sep-Pak Light C_{18} cartridge contains ~130 mg of sorbent.

Note: Remember that there is no Light equivalent to the long body plus design, so that the internal volume difference is more significant since the Light design has ~1/6 the internal volume of a long body plus cartridge.

Therefore, if you have a Sep-Pak Plus long body Florisil™ cartridge, it will contain ~910 mg of sorbent. The Sep-Pak Light Florisil™ cartridge will contain ~145 mg of sorbent. When converting from larger to smaller, or smaller to larger cartridges, scaling of the sample volumes and solvent volumes will be required. Slower liquid flow rates will be needed in the Light Body to adjust for the smaller internal diameter which impacts the linear velocity. This is discussed in the "Key Terms and Calculations" section, page 66.

On-Line Processing

For on-line processing, which can also be referred to as "in-line" processing, specially designed devices have been created for connecting the SPE device directly into the higher pressure flow paths of your liquid chromatographic system. These devices can be constructed of stainless steel similar to analytical LC columns, see Figure 76.

Figure 76: On-line SPE Devices

Programming the control of the flow paths with proper valving will allow you to effectively run the steps in an SPE method automatically on the liquid chromatography [LC] system. These on-line SPE column designs are capable of being re-used up to hundreds of times depending on the complexity of the sample matrix. The LC instrument will properly load, wash, elute, and re-generate the column for the next sample. Figures 77 and 78 show how the controlled switching of the valves will provide the correct flow and direction of conditioning and equilibration solvents as well as the loading, wash, and elution solvents for each step.

[On-Line Processing]

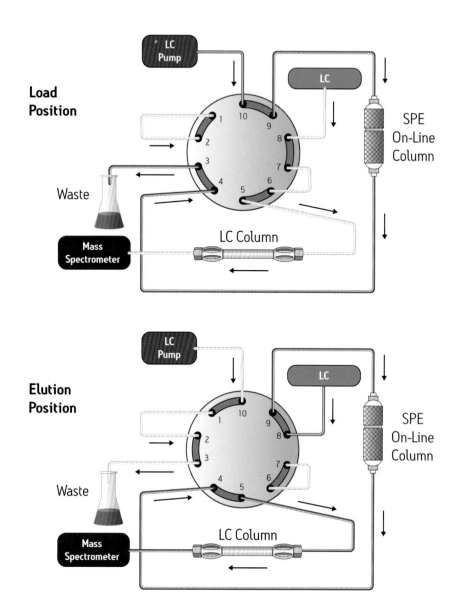

Figure 77: Flow Diagrams of an On-line SPE System During the Load and Elution Steps

Multiple SPE devices can be processed in an on-line system which provides the advantages of greatly reducing analysis time. For example, one device can be loaded while the other device is eluted and then regenerated for another sample. See Figure 78. If sensitivity is a concern, the SPE cartridge can be loaded and washed in one direction and eluted in the reverse direction [back flush]. This allows the analyte band to remain more concentrated.

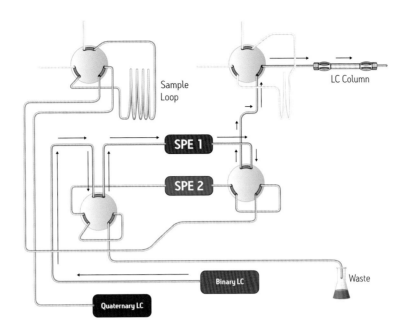

Figure 78: A Complex On-line SPE System with 2 SPE Devices

Please note that there are commercially available on-line technologies designed to use disposable single-use cartridges in specially designed instrumentation. These are not discussed in this book.

Dispersive SPE

Another format for performing SPE-based sample preparation, especially useful for screening purposes, is dispersive solid-phase extraction [d-SPE]. This technique utilizes sorbents and buffers which are in a powder or salt form and used loose in conjunction with tubes. While these types of methods have existed for many years, they have most recently gained wider attention with the development of QuEChERS [an acronym for Quick, Easy, Cheap, Effective, Rugged and Safe] methods. These are particularly useful if you want to screen for the presence of pesticides in a wide range of different types of food sample matrices and want to use the same, simple protocol. The driving force to move liquids in d-SPE methods is a shaking step combined with centrifugation. Waters DisQuE™ dispersive sample preparation products provide the capability and are shown in Figures 79–81.

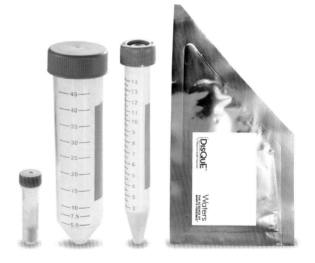

Figure 79: DisQuE Dispersive Sample Preparation for QuEChERS

[Dispersive SPE]

The different complex food sample matrices are processed using two d-SPE tubes, following a simple standardized protocol. No method development is required when changing to different types of samples. From example, the pesticide screening protocol for avocados is the same as for watermelons. Simple solvent additions and centrifugation are the only variations needed.

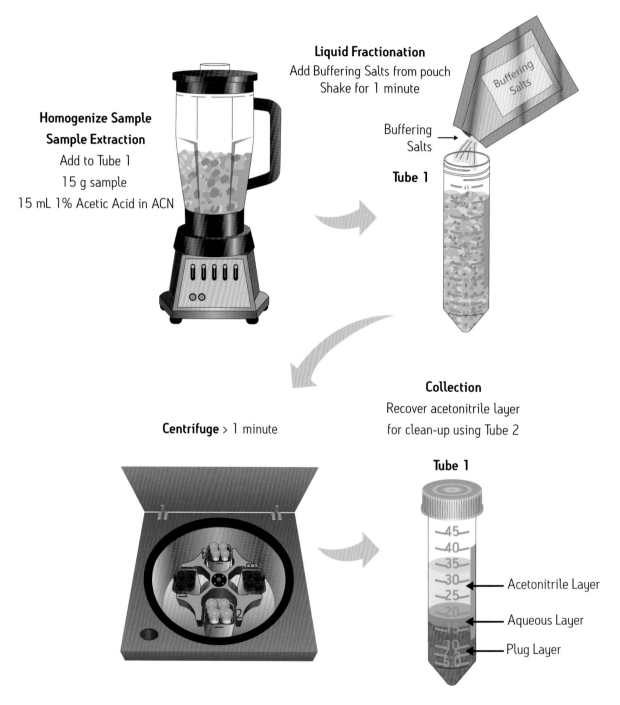

Figure 80: DisQuE Tube 1 Protocol [QuEChERS]

[Dispersive SPE]

The benefit of this technique is that the same protocol and devices will provide excellent fast screening results for pesticides as well as other contamination in samples of many different food matrices.

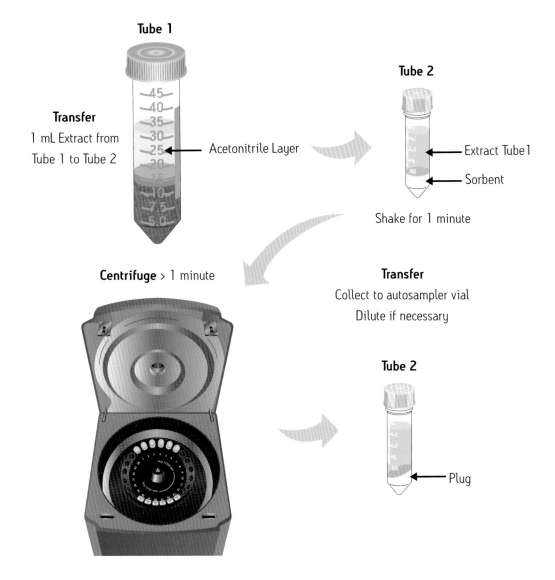

Figure 81: DisQuE Tube 2 Protocol [QuEChERS]

[Driving Forces—Gravity]

Basic Set Up-Driving Forces Used to Generate Liquid Flow

In order to process your sample, you have to achieve the desired flow of the liquid sample matrix and the solvents through the devices. Since this technology is based on LC, controlling the flow rate of the liquids is important in order to achieve the best performance. It is especially important not to flow too fast while the sample matrix is inside the cartridge, as this will decrease chromatographic efficiency. Another important point is that when you stop the liquid flow in the cartridge, there will still be liquid present internally in the sorbent bed, as well as inside the outlet tip. Remember to account for this liquid when you develop your method. See page 62.

Let's discuss the three different driving forces commonly used: gravity, positive displacement [pumping], and vacuum.

Gravity

Most SPE devices contain sorbents that range in particle size between 50 and 150 μm, which allows gravity to generate an adequate flow for most liquids and solvents through the devices. Figure 82 shows several different setups. The first two setups, starting left to right, use cartridges that have an available open volume above the packed bed of sorbent. The straight barrel syringe configuration [labeled VAC], and the enlarged opening configuration [labeled VAC RC], can have sample matrix and solvents added directly as the volumes permit. Note that in some cases, if the amount of liquid exceeds the open volume, then multiple, partial volume additions can be made. Depending on the number of samples to be processed, you can set up multiple cartridges and process several at a time in parallel. The only downside of using gravity is that the flow rate is usually very slow, which increases the time it takes to process a sample.

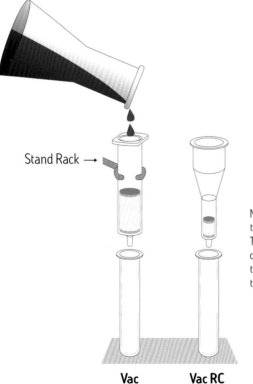

NOTE: There are many different ways to apply a sample onto an SPE cartridge. Typically pipettes or adapters with tubing can be used. This flask will represent the generic ways liquids can be applied to the cartridges.

Figure 82: Using Gravity as a Driving Force [also see Figure 93 on page 97]

[Drivnig Forces—Centrifugation]

The three new set ups, pictured on the right side of Figure 83, show the enclosed cartridge. For enclosed devices where there is no available open volume, such as Classic, Plus, and Light designs, use the inlet female Luer connections. A reservoir with a male Luer outlet is connected to the inlet of the SPE device. The sample and liquids can be added with gravity controlling the flow rate.

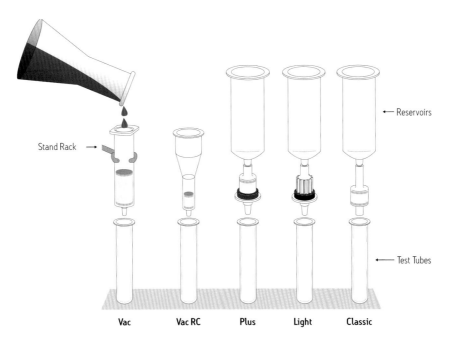

Figure 83: Using Gravity as a Driving Force

As an alternative solution, centrifugal force can be used to create increased, artificial gravity by placing these cartridges into properly configured and balanced centrifuges for controlled processing by spinning. This will reduce the amount of time required, and in addition allows you to remove the solvent completely from the sorbent bed, if you desire. DisQuE products, discussed previously, utilize centrifugation to process the samples. See Figure 84.

Figure 84: Using Centrifugal Force

[Driving Force—Positive Displacement or Pump Pressure]

Positive Displacement or Pump Pressure

The second driving force is using positive displacement, or pumping, of the sample and liquids through the cartridge. There are many different ways to achieve flow using this driving force. Figure 85 shows a very convenient technique if you do not have a large number of samples to process. A simple syringe/plunger connected to the configurations with female Luer connectors [Plus, Light, Classic] works well.

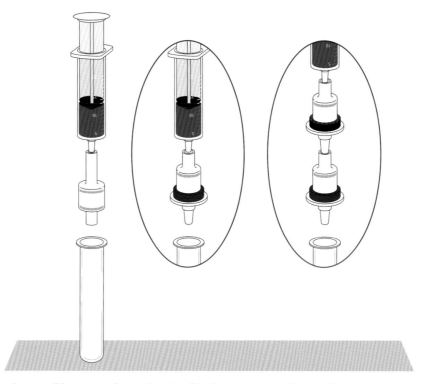

Figure 85: Using a Syringe/Plunger to Create Positive Displacement as a Driving Force

The flow rate and processing time will be controlled by how hard you push the plunger. Watch that the sample and liquids flow slowly enough so that drops are formed at the outlet of the cartridge [~1–2 mL/min.], as shown in Figure 86.

If you push too hard, then a steady flow stream will be created, indicating that you are flowing too fast. This will result in poor chromatographic performance in the cartridge, and poor processing of your sample. On the left of Figure 87, the green analyte is fully retained when the sample is loaded on the cartridge at the proper flow rate. The diagram on the right shows some of the green analyte being flushed out of the cartridge since it is being loaded with too fast a flow rate.

Figure 86: Proper Flow Rate Control–Drops

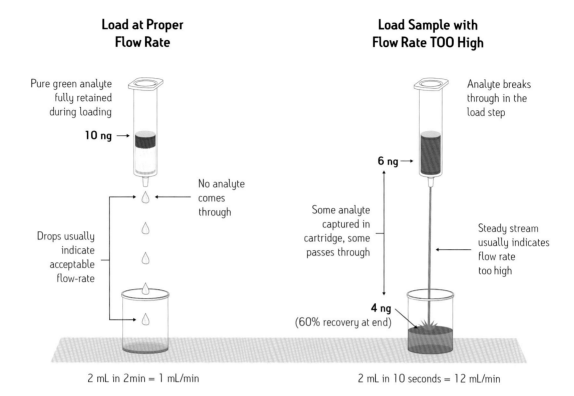

Figure 87: Importance of Flow Rate Control

Syringe barrel cartridges can also be configured for positive displacement by the use of an adapter that fits the inlet end of the barrel, and has a female Luer opening that can be connected with a pumping device, such as a syringe with plunger, to create the positive displacement flow [Figure 88] The open volume of the SPE cartridge above the sorbent bed must be taken into account in the method development.

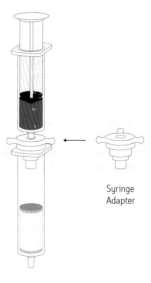

Figure 88: Adapting a Syringe for Positive Displacement

[Driving Force—Positive Pressure]

A true pump can also be used, as shown in Figure 89. Here, male-male adaptors are used to connect the cartridges to the tubing. Again, watch for the formation of drops to indicate proper flow rate. Pumps that can be set for a constant flow rate will offer more consistent results, sample to sample.

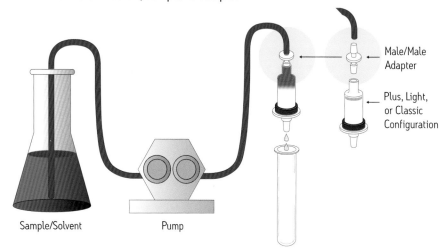

Figure 89: Adapting an Enclosed Cartridge for a Peristaltic Pump

If a large number of samples must be processed, then investigate various SPE instrument suppliers for equipment that can control the processing of several cartridges in parallel using positive displacement. When processing a large number of samples some instruments can be programmed to automatically execute all the steps of your SPE method. The 96-well plate format is easily handled by most automated equipment. All the throughput benefits of the 96-well plate design can be optimized with the added precision of constant flow rate for each of the 96 wells.

For faster processing of the plate format, without the need of true automation, then a positive pressure manifold can be used. See Figure 90.

Figure 90: Positive Pressure Processor for 96-Well Plates

Vacuum

Vacuum is commonly used to achieve liquid flow in SPE cartridges. This is popular for applications where a moderate to large number of samples must be processed, and the equipment cost needs to be minimized. Vacuum manifolds are relatively inexpensive, and designed specifically for this type of processing. The manifold design, shown in Figure 91, is intended for individual SPE cartridges, where up to 20 different samples can be processed in parallel.

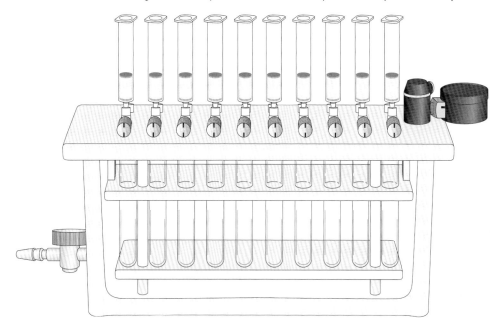

Figure 91: Glass Vacuum Manifold

The cartridges are connected on the top manifold cover, which has 20 female Luer connector positions available. These connectors have a valve assembly [known as a stop-cock valve] that is used to control the liquid flow rates of each SPE cartridge, and to seal any openings if not all 20 positions are being used. The manifold cover has a seal to mate with the top edge of the glass vessel. The liquids and samples can be pipetted directly into the open volume of the syringe cartridges. A vacuum source is applied to the connector at the lower left of the glass vessel. This can be a vacuum pump or a water aspirator vacuum source. The applied vacuum in the glass vessel pulls the liquids down through the cartridges. A construction strength glass vessel is usually used so that you can monitor the flow from each cartridge [i.e. drops], and the glass is also impervious to solvents.

Depending on the step of the SPE method being executed, different collection containers can be used. For example, waste reservoirs [for the conditioning, equilibration, and wash steps] and test tubes or vials can be fitted in racks, that are located in the glass vessel, to capture analytes at the outlet of each SPE cartridge.

Note: Flow rate control, as discussed above, is critical even if you use vacuum as the driving force, since a strong vacuum can create flow rates that are too high, resulting in poor performance.

[Driving Force—Vacuum]

This type of vacuum manifold can be used in many different ways depending on your application and choice of SPE cartridge configuration. Figure 92 shows some of this operating flexibility. Note that all the various individual cartridge configurations can be connected to the vacuum manifold.

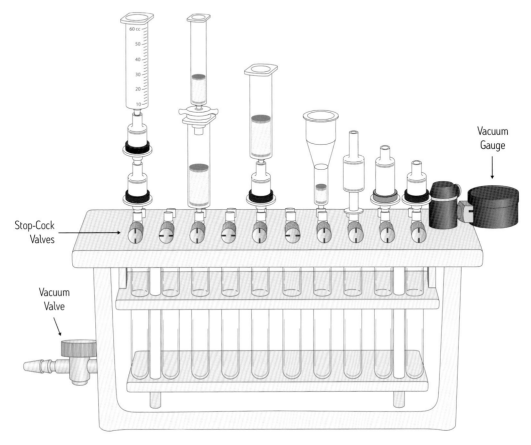

Figure 92: Set up of Glass Vacuum Manifold

You can see that the male Luer outlet tips of the Plus, VAC, and VAC RC designs mate directly with the female Luer stop-cock valve connector. Located third from the right, the Classic design, with a female Luer outlet, can be connected to the stop cock valve using a male-male adapter.

Applying Samples and Solvents to Cartridges

Now, let's look at how liquid samples and solvents can be applied to cartridges. Usually, the vacuum is turned OFF, as the liquids are applied to the cartridges. This is important in order to provide more consistent processing of all the cartridges. For example, when you have twenty cartridges ready to process, you will want to control the flow rate and time the liquids [samples and solvents] are in the cartridge. If the vacuum is ON when you load the first cartridge, then all the liquid could flow through that cartridge before you finished loading all the other cartridges. The first cartridge will now have a significant amount of air being pulled through it, because its liquid is gone, and the vacuum is still ON. This can cause poor reproducibility when comparing the results from the twenty samples. So, it is best to turn the vacuum OFF until all the cartridges have had their samples or solvents applied.

In the following diagrams, you can see how the syringe barrel design can have liquids applied in different ways.

In the first example in Figure 93, the liquid volume to be applied is less than the open volume above the sorbent bed [for example, loading 1 mL of solvent into a cartridge with an open volume of 2 mL]. Pipette the specified volume of liquid into the cartridge and then apply the vacuum when all the cartridges are filled. Use the stop-cock valves to keep the flow rates consistent for all the cartridges. Pouring of liquids is usually not recommended unless correct volumes can be assured.

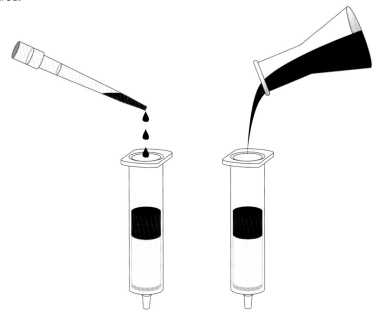

Figure 93: Applying Liquids to Syringe Design

The second method, Figure 94 is used when the amount of liquid is somewhat larger than the open volume—for example, if you have 15 mL of solvent to load onto the SPE cartridge with an open volume of only 5 mL. As shown, a reusable, syringe barrel adapter is available that allows you to connect an empty syringe reservoir above the inlet of the SPE cartridge. Then, after all the reservoirs are loaded, apply the vacuum so that they all flow together. Use the stop-cock valves to keep the flow rates consistent for all the cartridges and remember to limit any air from being drawn into the cartridges as this can cause variable analytical results.

[Applying Samples and Solvents to Cartridges]

The third technique, Figure 95, is used especially when you have a very large sample volume, such as a liter of drinking water, for which you need to perform trace concentration in order to quantify the low level contaminants present. The sample can be stored in a large bottle and, by using a male-male adapter and a syringe barrel adapter, you can connect a piece of clean, compatible tubing from the bottle to the inlet of the SPE cartridge. Apply the vacuum to begin flow, which is controlled by the stop-cock valve. Proper collection containers will be necessary in the manifold. The analytes are captured by the cartridge and the large liquid volume that passes through the cartridges can be drawn out at the bottom vacuum line connector of the glass vessel into a collection trap. Use the stop-cock valves to keep the flow rates consistent for all the cartridges, and remember to limit any air from being drawn into the cartridges as this can cause variable analytical results.

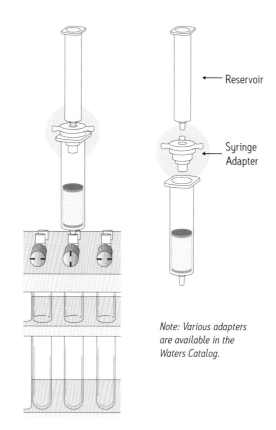

Note: Various adapters are available in the Waters Catalog.

Figure 94: Adapting Resevoirs

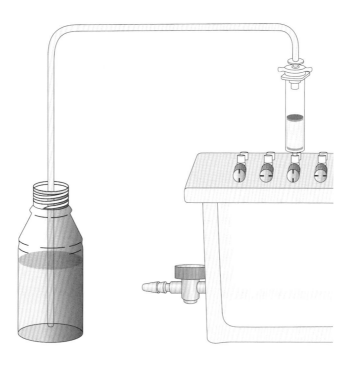

Figure 95: Applying a Large Volume Sample

[**Using Vacuum with Enclosed Cartridges**]

Using Vacuum with Enclosed Cartridges

When using vacuum with enclosed cartridge designs [Plus, Light, and Classic], refer to Figure 96 to see how to set-up the cartridges. Note that the Classic design will require a male-male adaptor to connect with the stop-cock valve. Also, the Light design is not shown.

There are two primary ways to charge liquids into these cartridges, which feature female Luer inlet connectors.

For larger volumes, you can draw the sample volume through tubing with a male/male adaptor connected to the cartridge as shown on the left of Figure 97. In all cases, it is important to record the exact amount of sample volume that has been processed through the cartridge so that you can properly calculate the method's performance at completion.

For smaller volumes, mate an empty syringe barrel body with a male Luer outlet as a reservoir, as shown on the right of Figure 97. Fill all the reservoirs for each cartridge, and then apply the vacuum. Control individual cartridge flow with the stop-cock valves, and remember to limit any air from being drawn into the cartridges as this can cause variable analytical results.

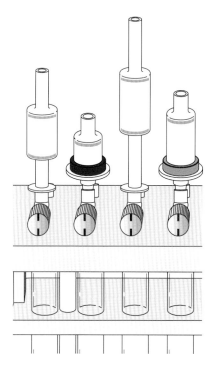

Figure 96: Enclosed SPE Devices

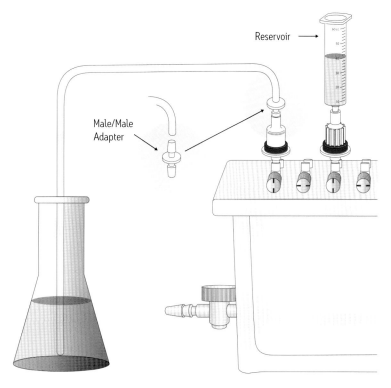

Figure 97: Applying Liquids to Enclosed SPE Devices

[Using Vacuum with 96-Well Plates]

Using Vacuum with 96-Well Plates

Special vacuum manifolds are designed for the 96-well plate configuration, as shown in Figure 98.

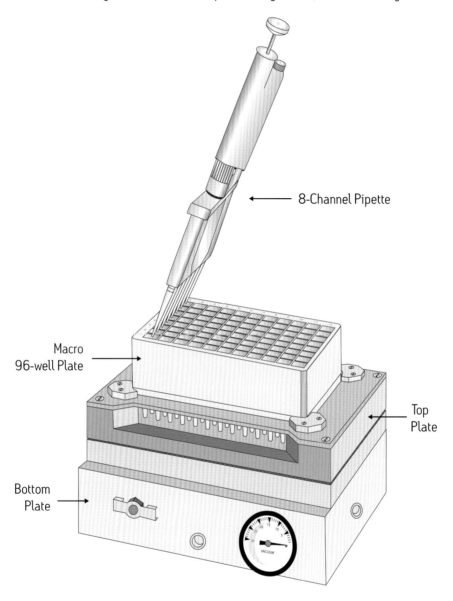

Figure 98: 96-Well Plate on Manifold

Here, the liquids are charged into the open volume above the sorbent bed in each well, while the plate is placed on top of the manifold. Typically, an eight-position pipetter [see Figure 98] is used to fill eight wells simultaneously to save time. If the liquid volume is greater than the open volume above the sorbent bed, then multiple additions of the liquid will be required. Special 96-position collection plates and waste reservoirs are used with the vacuum manifolds. Once all wells are filled, then vacuum can be applied. Since 96-well plates have no stop-cock valve to control the individual flow rate of the wells, care must be taken in order to obtain consistent analytical performance especially when silica based reversed-phase sorbents are used. The Oasis family of SPE sorbents was especially designed to work successfully in the 96-well plate format without the need of stop-cocks to control flow during the conditioning and equilibration steps.

Steps in an SPE Method

An SPE method, or protocol, consists of a series of steps that are followed in sequence. In this section, we will provide an overview of the name, and function of each step. A typical SPE method is shown in the box diagram below. Please note that in some SPE methods, not all of these steps will be followed. Let's describe the importance of each step.

Generic SPE Method/Protocol

- Pretreatment of Sample
- Conditioning of Cartridge
- Equilibration of Cartridge
- Load Sample
- Wash Away Interferences
- Elute Compound[s] of Interest
- Dry Down and Reconstitute Compounds in Mobile Phase

Pretreatment of the Original Sample

In some methods, this step is not necessary. However, it will be necessary in the following, common situations:

a. If the sample is a solid, dissolve in an appropriate solvent, or homogenize the solid followed by an extraction of the analyte[s] of interest into a solvent. In both cases, if this solvent is too strong for the type of chromatography to be performed in the SPE cartridge, then a dilution will be required.

b. If your compounds of interest are ionizable, and will encounter an ion-exchange sorbent in the cartridge, then adjust the sample pH to put them into the appropriate ionization state.

[Preparing the Cartridge]

c. If your compounds of interest are in biological fluids, and are naturally bound to the proteins present, then they must be released from the proteins before entering the SPE cartridge. Usually, an acidification or basification of the sample will be sufficient.

d. For proper analytical quantitation, the addition of an internal standard to the sample may be required.

Preparing the Cartridge

Once the sample has been properly pretreated, we can proceed with the remaining steps of the protocol. To help understand each step, we will use the following sample preparation situation. For this protocol, a reversed-phase sorbent SPE cartridge is being used to process an aqueous "green" sample matrix. Our goal will be to remove a "yellow" interference from our "blue" analyte of interest. The chromatographic behavior of the sorbent bed will initially retain both of these compounds. In SPE, when compounds are retained, this is called "capturing" of the compounds. Then, we will selectively separate and wash the yellow interference from the blue analyte using a stronger mobile phase solvent. Finally, we will release the purified blue analyte from the cartridge with an even stronger solvent. The conditioning and equilibration steps are used to prepare a cartridge for the sample.

However, before we load our sample onto the SPE device, we must check for the type of sorbent being used. This will determine if the sample can be applied directly to a dry cartridge, or if the cartridge must be wetted before the sample is applied. Use the guideline in Table 8.

Table 8: Requirements for Condition and Equilibration Steps

Step	Normal Phase	Reversed Phase
Condition	Not necessary	Always required
Equilibration	Suggested	Always required

Before the sample is loaded, the wetting of the pores in a reversed-phase sorbent is achieved by the conditioning step in combination with the equilibration step.

Condition the SPE Cartridge

SPE cartridges contain sorbents that are very porous [similar to the pores of a sponge] in order to provide a large chromatographic surface area to create the separations. This large chromatographic surface area is contained inside the sorbent pores, and is responsible for the performance of the cartridge. In order for the compounds to be chromatographically acted upon by the sorbent, they must be able to flow into and out of the sorbent pores. This is achieved when the sorbent pores are properly wetted by the sample solvent [mobile phase]. See Figure 99.

[**Condition the SPE Cartridge**]

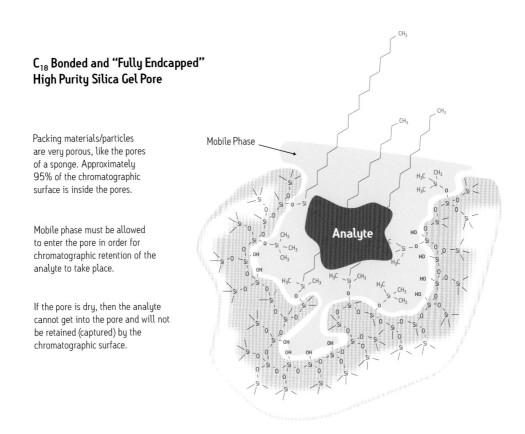

C$_{18}$ Bonded and "Fully Endcapped" High Purity Silica Gel Pore

Packing materials/particles are very porous, like the pores of a sponge. Approximately 95% of the chromatographic surface is inside the pores.

Mobile phase must be allowed to enter the pore in order for chromatographic retention of the analyte to take place.

If the pore is dry, then the analyte cannot get into the pore and will not be retained (captured) by the chromatographic surface.

Figure 99: Sorbent Pores must be Properly Wetted

As mentioned above, the need for the conditioning step will depend upon the type of sorbent and the mode of chromatography to be performed in the cartridge. In general, for normal-phase chromatography [such as unbonded silica, Florisil, and alumina sorbent particles, the sample will be in an organic sample solvent, which can provide the proper wetting. Therefore, the conditioning step is not necessary. For more information on normal-phase chromatography, see page 32.

However, if reversed-phase chromatography is used, the sorbent must be properly solvated, or wetted before processing can begin. Under reversed-phase conditions, the sorbent such as a C_8, C_{18}, and phenyl particles, etc., are chemically bonded to silica, making them hydrophobic. An aqueous sample is not compatible with this hydrophobic surface because the sample solvent will not enter the dry pores: therefore, the compounds of interest will not be able to interact with the chromatographic surface area.

Since our protocol for the "green" sample uses reversed-phase conditions, we must first wet the sorbent surface. A strong organic solvent, such as methanol, acetonitrile, or isopropanol is used. This solvent properly wets the very porous surface of the sorbent particles. The volume of solvent required will depend upon the amount of sorbent contained in the cartridge. See Figure 100.

[Equilibration]

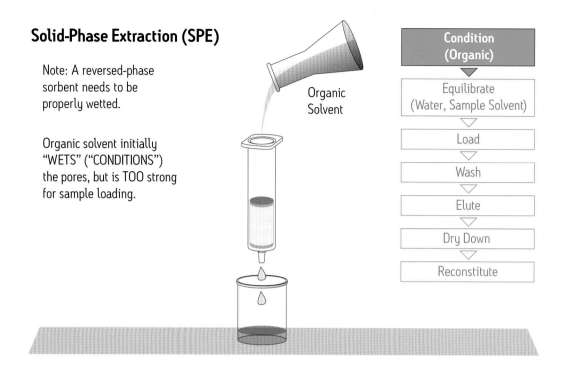

Figure 100: Conditioning the Pores of a Reversed-Phase Sorbent with Organic Solvent

Once wetted, the cartridge must then be made ready to accept the aqueous sample by performing the equilibration step.

Equilibrate the SPE Cartridge

It is important to equilibrate the sorbent with the same solvent as the sample matrix. This ensures that the solvent is not too strong for the sample, which would otherwise cause poor chromatography. For example, in normal-phase chromatography, the sample will typically be in an organic solvent. Therefore, pass the same organic solvent that the sample is in, through the cartridge to ensure proper equilibration. However, in some normal-phase protocols, the sample maybe loaded directly onto a dry cartridge.

For reversed-phase chromatography, the sample is typically in an aqueous sample solvent, which is much weaker than the conditioning solvent that is now in the pores. You will need water or a weaker sample solvent to be equilibrated onto the surface of the sorbent in the pores, before the sample can be loaded. See discussion on mobile phase elution strength on page 30.

Remember that a reversed-phase C_{18} sorbent is very hydrophobic so that it does not want to interact with water or aqueous solvents. That is why the conditioning step, discussed in the previous section, is performed first. The C_{18} is already wetted with the strong organic solvent. That solvent is miscible with the water, so now we can pass water through the cartridge to replace the organic solvent in the pores with water. The replacement of the organic solvent now allows the sorbent pores to be wet with water. This will achieve an equilibration with the aqueous sample which can then be applied, and the compounds will be able to interact in the chromatographic pores. If the sample was in some other aqueous sample solvent, use that sample solvent for the equilibration step. See Figure 101.

[Equilibration]

Without this equilibration step, the full chromatographic performance of the sorbent bed cannot be obtained. The volume of equilibration solvent will depend on how much sorbent is in the SPE cartridge. An SPE cartridge that has not been properly equilibrated will not be able to retain the expected sample compounds for the sample matrix that is applied to that specific cartridge. This will cause poor analytical results for that sample, resulting in the need to reprocess that sample.

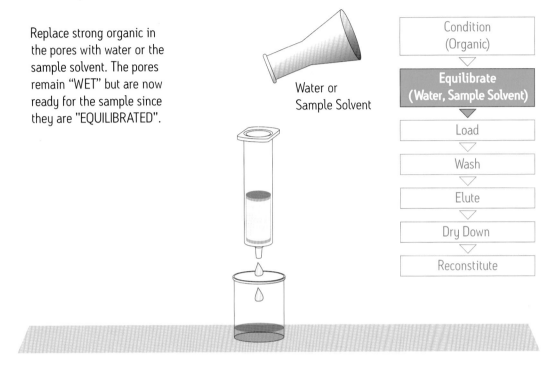

Figure 101: Equilibrating the Pores with Water or Sample Solvent

Note: During the conditioning and equilibration steps, the sorbent bed should never be allowed to run dry before the sample is applied. This will cause the pores to dewet.

For example, if the conditioning and equilibration steps were not done properly and if some air got into the cartridge, the compounds, which should be captured, will pass directly through the cartridge resulting in a poor clean-up of the sample. This can happen, if a vacuum manifold is used and some of the twenty cartridges did not get properly conditioned and equilibrated. [This can also happen when 96-well plates are used with vacuum.] The results from those improperly wetted cartridges will require these samples to be reprocessed. The control of the flow rate during the conditioning and equilibration steps does not have to be monitored closely, since the sample has not been applied yet. However, the sorbent bed should never be allowed to run dry, with just air flowing through, before the sample is applied.

In Figure 102, the cartridge on the left was properly conditioned and equilibrated. The green sample was a mixture of yellow and blue dyes. Both dyes were well retained [captured] at the inlet of the cartridge on the properly wetted sorbent bed. The cartridge on the right was not properly wetted. When the sorbent pores are de-wetted, the two dyes are not captured as they were supposed to be moving through the cartridge during the loading step. Even though, the yellow and blue dyes were partially separating as they were loaded onto the cartridge, this sample would have to be retested to get the desired results.

[Load the Sample]

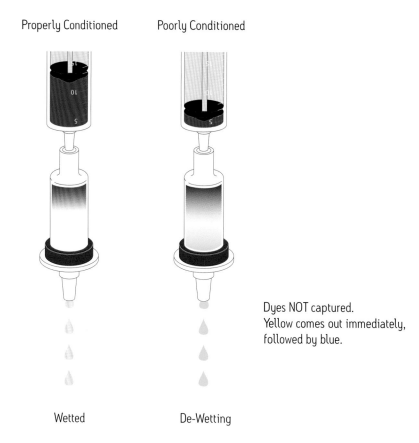

Figure 102: Proper Wetting of Pores is Critical for Capture During Loading Steps

Load the Sample

When the actual sample is applied to the cartridge is called the loading step. The sample contains the analyte[s] or compound[s] of interest as well as the other sample matrix substances.

As a part of the method development process, the total amount of sample to be loaded will have to be determined. It will be dependent upon the amount and type of sorbent in the SPE cartridge, the type of sample solvent, and the exact sample matrix. For example, if you develop a SPE method for an analyte [compound] in dog plasma, and there is a need to change to human plasma, then the method will have to be re-evaluated, since the sample matrix, contains different substances. The new substances present in the matrix can result in poor SPE performance. The amount of sample [volume] will be specified for the sorbent type and size of the cartridge.

"Determining the Proper Load Capacity" in the Methods Development section will detail how to select the size of the SPE cartridge for the volume of sample to be processed. See page 154. The actual mass of the compound[s] of interest loaded onto the SPE device will be determined at this step from their concentration and the volume of sample. This value will be used later to calculate the performance of the protocol. Since chromatography is taking place as the sample is being loaded onto the cartridge, it is important to monitor the flow rate of the liquid exiting the cartridge.

[Load the Sample]

Chromatographically speaking, the flow rate is related to the linear velocity of the liquid, which is the speed of the liquids in the SPE cartridge. It is important to control the liquid speeds in order to ensure proper performance. Slower flow rates are needed in order to allow the chromatography to work properly. As a guideline, "liquid drops" should be seen exiting the cartridge, indicating a flow rate of approximately one to five mL/min. A steady stream of liquid is too fast, unless this faster flow rate was studied during the method development process and proven to have no effect on the chromatographic process.

Remember that flow rate control is important during the load, wash, and elution steps. For more information, see "Flow Rate Control" in the "Key Terms and Calculations" section on page 65. The speed of the liquid is also impacted by the internal diameter of the SPE cartridge. When scaling to a different size cartridge, the linear velocity must be maintained, which means the flow rate must be adjusted. This is discussed in more detail in the section "Scaling to different size cartridges, and Linear Velocity and Flow Control of Liquids" in the Key Terms and Calculations section. See page 66.

We are now ready to load our sample, as shown in the Figure 103 below.

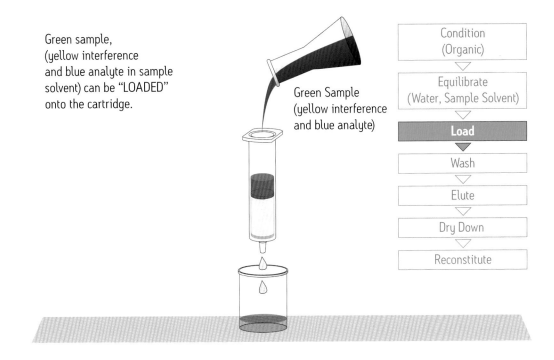

Figure 103: Load the Sample

Note that under the chromatographic conditions in the SPE cartridge during the load step, both the "yellow" interference substance and the "blue" analyte of interest are well captured. They both remain at the inlet of the sorbent bed because they are more attracted to the sorbent than the sample matrix solvent. They appear as a "green band" because the yellow and blue are still mixed together and retained, and therefore appear as "green" to our eyes.

[Wash Away Interference]

Wash Away Interference[s]

The wash step is important since it allows the sample matrix to be simplified, removing substances that will cause problems in the analysis of the compound[s] of interest. Within a typical sample matrix, some interfering substances have less retention than our compound[s] of interest, and some substances will be more strongly retained by their attraction to the sorbent. The wash step is designed to remove those substances with less retention, while leaving the compound/analyte of interest and other more strongly retained substances, still captured [retained] by the sorbent. The substances "washed" from the SPE cartridge are then discarded to waste, following appropriate safe guards. See Figure 104.

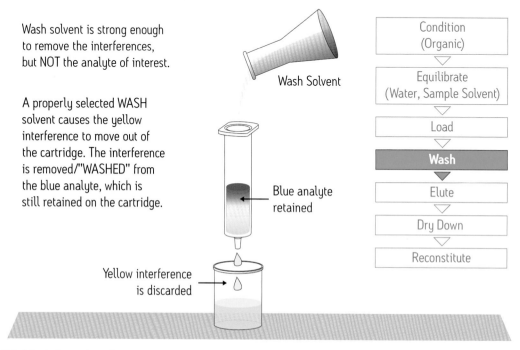

Figure 104: Wash Away Interferences

The "yellow" interference has less relative attraction to the sorbent than the "blue" analyte. Since they have different levels of chromatographic attraction to the sorbent, this can be used to create a powerful sample clean-up wash step. By carefully selecting a new solvent which is just strong enough to remove the less retained "yellow" interference but not strong enough to move the "blue" more strongly retained analyte, we can achieve our desired goal. The "yellow" interference will be washed from the cartridge, with the "blue' analyte still being retained by the sorbent. Determining the appropriate wash solvent and volume necessary to complete the separation, will be covered during the method development phase.

Chromatographically speaking, we have just separated out a subset of the sample matrix, represented by the "yellow" interference substance. This is called "fractionation." As mentioned earlier, this "yellow" fraction could represent a whole class of interfering substances with similar retentions that were all removed by this one SPE wash step. Since the wash solvent was able to remove these substances from the SPE cartridge, they were technically "eluted" from the chromatographic/sorbent bed. This step is called a "wash" step when the substances removed are to be discarded. If further analytical work was to be carried out on this fraction, then this step would also be called an "elution" step.

To summarize, the output from a "wash" step is discarded, and the output from an "elution" step goes on for further analysis. For very complex sample matrices, there will be some SPE methods with multiple wash steps.

Since chromatography is being performed during the wash step, control of the flow rate [linear velocity] must be maintained.

Elute the Compound[s] of Interest

Our SPE method now requires the use of a stronger solvent to elute the "blue" compound of interest that is still retained on the sorbent. Since we are removing the compound for further analysis, this is called an elution step. The specific solvent with the required strength to remove the compound is called the elution solvent. The compound[s] of interest and the elution solvent that are collected from the cartridge, are called the eluate. See Figure 105.

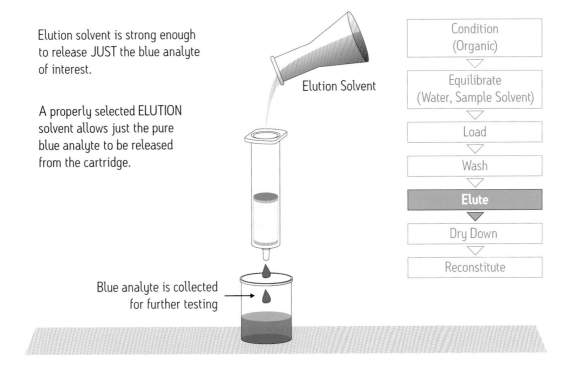

Figure 105: Elute the Compounds of Interest

For a more complex sample matrix, the goal of the elution step is to select the right solvent to remove only the compound of interest while leaving interfering substances that have a stronger attraction to the sorbent still retained inside the cartridge. Since they are of no analytical interest, these unwanted substances can be disposed of along with the cartridge following proper safety practices.

[Dry Down and Reconstitute in Mobile Phase]

In some protocols, there may be several compounds of interest with very different retention characteristics. There may also be a need to group certain compounds together in similar classes for separate, subsequent analytical testing. An SPE method would be developed to include multiple elution steps, with each elution step utilizing a different strength solvent. This is called fractionation and uses a step gradient of increasingly strong solvents to separate the compounds of interest. Fractionation of a sample using SPE can be a very powerful way to improve the overall analytical efficiency in the laboratory. This will be discussed in the Methods Development section. See page 125.

Dry Down and Reconstitute in Mobile Phase

The need for the drying down and reconstitution steps is dependent upon the sample condition requirements of the subsequent analytical technology. In many SPE methods, this step is not necessary because the eluate from the SPE cartridge can be analyzed as is or with a simple dilution. However, since a strong solvent was used to elute the analyte from the SPE cartridge, this strong solvent can create problems in some types of analytical techniques which follow.

Certain analytical technologies will require that the eluate go through two modifications. First, a dry down step is performed, usually with a flow of nitrogen. This evaporates just the strong solvent that was needed to elute the analytes from the SPE cartridge, leaving behind just the analytes of interest. Then, a weaker solvent, or the mobile phase used in the liquid chromatographic application, is added to the analytes. This makes the prepared sample more compatible with the analytical instrumentation. This is called "reconstitution". See Figure 106.

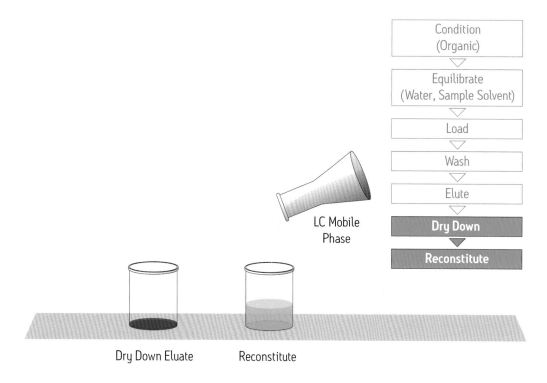

Figure 106: Dry Down and Reconstitute in LC Mobile Phase

[Dry Down and Reconstitute in Mobile Phase]

A good example of this situation is when reversed-phase LC is to be performed on a sample that had been processed using reversed-phase SPE. In reversed-phase LC, the sample solvent must be weaker than the mobile phase running in the LC system. If the sample solvent is too strong, then poor chromatography, with poor peak shapes, occurs. Look at the upper chromatogram in Figure 107. Note the poor peak shape for the three analytes. This was caused by a too strong sample solvent, which was 80% organic, being injected into the LC running a weaker mobile phase of only 40% organic.

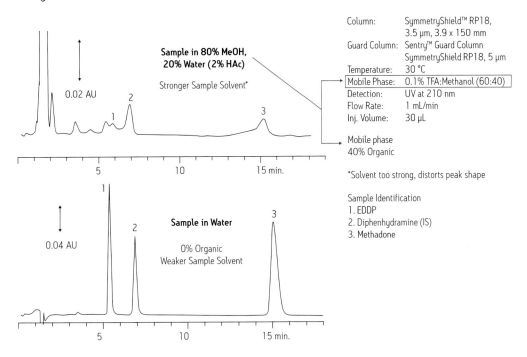

Figure 107: Why Dry Down and Reconstitution Steps are Required

In the lower chromatogram, the exact same LC conditions are being run, but now note the excellent peak shapes. The difference was that the strong 80% organic sample solvent was evaporated and the three analytes were reconstituted in pure water, which is a very weak solvent in reversed-phase chromatography. This was then injected into the LC system, and the chromatography was greatly improved for the three analytes.

In summary, the number of steps needed for an SPE method will be dependent upon the complexity of the original sample matrix, the specific goals for the SPE process, the mode of chromatography [i.e. normal phase, reverse phase, ion exchange], and the type of subsequent analytical technology to be performed.

Method Development

DEVELOPING YOUR SPE METHOD

There are many approaches to solid-phase extraction [SPE] method development. However, it is recommended that you first begin this process by writing down your goals. As you draft the goals and specific requirements for your SPE protocol, some of the issues you should consider are:

- Whether a validated method is needed, or if the method will only be used for a short study
- Concentration and sensitivity requirements for compounds of interest [ppb, ppt, etc...] and detection scheme
- Characteristics of compounds of interest [polarity, ionizable (pKa), temperature or light labile, etc...]
- Sample matrix and range of variation
- Complexity/nature of interfering substances
- Numbers of samples to be processed, also referred to as "Sample Throughput"

Use existing information as much as possible to help make the process go more efficiently. SPE technology may be new to you, but many scientists have used SPE to solve problems for over 30 years. You may be able to build upon their experiences. Several sources of information are readily available, including:

- Experienced colleagues
- Literature references to the exact or similar sample preparation problem
  - Google Scholar
  - Google Search
- SPE product vendors
  - Technical support
  - Applications databases
- Training organizations

Time invested in this upfront work will usually provide a significant reward in the successful completion of the project.

Sample/Analyte Attributes

Obviously, the more information you have on the nature of the analyte[s] of interest, the better equipped you will be for creating the best possible sample preparation method. It's important to understand details of the molecule, such as:

- Polar or non-polar nature [log P]
- Ionizable [pKa or pKb]
- Strong acid or base
- Zwitterionic
- Degree of hydrophobicity

If you do not have specific information on the analyte, Waters has method development protocols and devices designed to help you create a successful method without the need for detailed analyte information up front. A complete review of this section will provide key information for your method development effort.

Sample Matrix Pretreatment

In many methods, a pretreatment step is not necessary. However, it will be necessary in the following, common situations:

1. If the sample is a solid, then dissolve it in an appropriate solvent or homogenize it, fthen extract of the analyte[s] interest into a solvent. In both cases, if the extraction solvent is too strong for the type of chromatography to be performed in the SPE cartridge, then a dilution will be required.

2. If your compounds of interest are ionizable and will interact with an ion-exchange sorbent in the cartridge, then adjust the sample pH to put them into the appropriate ionization state.

3. If your compounds of interest are in biological fluids, and are naturally bound to the proteins present, then they must be released from the proteins before entering the SPE cartridge. An acidification or basification of the sample will usually be sufficient.

4. For proper analytical quantitation, the addition of an internal standard to the sample may be required.

Samples/sample matrices come from a wide variety of sources. In addition, the types and complexity of the compounds that are present can interfere with the analysis of analytes, resulting in significant challenges for the analytical scientist. References on how to best process a given sample matrix can be found in the Waters Applications Database found on the Waters Home page by typing in "Applications" in the search box or in web searches. Some are quite creative and use very unique ways of employing chromatographic devices. These different strategies are described below.

SPE Strategies to Solve Different Problems

Using a chromatographic sorbent bed to perform sample preparation in SPE is very powerful. The creativity of the analytical scientist, coupled with this powerful SPE "tool," can solve difficult analytical problems presented by many samples. This section discusses four distinct ways to use SPE devices for sample preparation problem-solving.

There are four different SPE strategies designed to accomplish different goals for the sample preparation process. These strategies can be described based on what you wish to do initially with the compound[s] or analyte[s] of interest:

1. Pass through of the analyte[s] of interest, thereby capturing interferences
2. Capture [retain] the analyte[s] of interest, washing away interferences
3. Capture and fractionate the analytes of interest with separate elution steps
4. Capture and trace- enrich [concentrate] the analytes[s] of interest, by processing of large volumes [100–1,000 mL] of a sample containing a very low analyte concentration level.

By manipulating the chromatographic conditions in the SPE cartridge [sorbent type and solvent choices], the analytical scientist can select the best, lowest cost strategy to solve the specific sample preparation problem. This is where the creativity of the method development scientist, combined with chromatography, can become very powerful in solving analytical problems.

Included in the discussion of these strategies will be diagrams showing you what is happening. Various cartridge designs will be used so that you can see how to manipulate those designs.

Strategy 1: Pass Through

This strategy is used when the sample matrix contains interfering substances that must be separated away from the analyte/compound[s] of interest. For this strategy to work, the analyte[s] themselves must be present at sufficient concentration levels for easy detection during the final analysis, since they pass through the SPE device without any further concentration.

The analyst selects the appropriate type of sorbent and solvent conditions so that the compound[s] of interest are NOT retained by the sorbent. However, all the interference substances ARE strongly retained by the sorbent. As the sample matrix is loaded onto the SPE cartridge, the compounds of interest can pass through the cartridge, while the interfering substances are captured by the sorbent. See Figure 108 . Here the "purple color" represents the analyte [compound] of interest and the interfering substances in the sample matrix are "colorless."

[SPE Strategy 1—Pass Through]

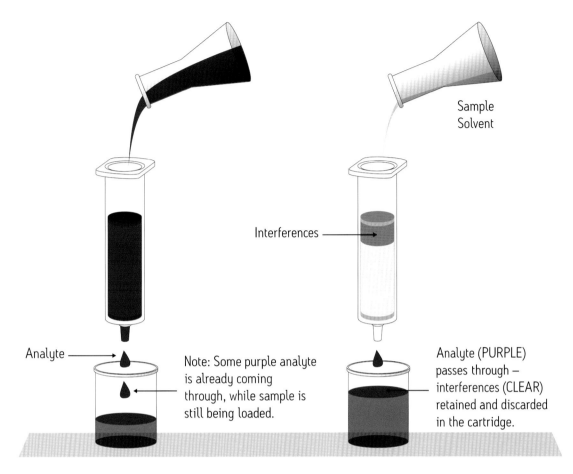

Figure 108: Strategy 1: Pass Through

As the sample matrix is being loaded, there is no attraction for the purple analyte to the sorbent because of the chromatographic conditions chosen by the analyst [sorbent type and sample solvent type]. Since chromatographic value k for that analyte is very close to zero, it moves freely through the sorbent bed. However, the k values for the interferences are high, so that they all remain captured, and can be disposed of with the used cartridge after processing.

Depending on the volume of sample and the hold-up volume of the SPE cartridge, two situations will occur.

1. Sample Volume Greater Than Hold-Up Volume

First, if the sample volume is greater than the cartridge hold-up volume, then some of the purple analyte will actually exit from the cartridge during the loading step and must be collected. Note that as the sample volume is loaded onto the cartridge, it will replace that same amount of equilibration solvent that remained in the cartridge from the previous step. This is typically collected along with the initial portion of the purple compound of interest that elutes during the load step. See Figure 109.

[SPE Strategy 1—Pass Through]

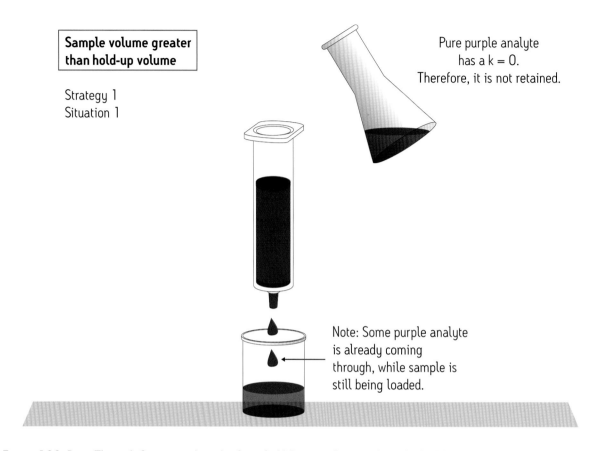

Figure 109: Pass-Through Strategy when the Sample Volume is Greater than the Hold-up Volume

However, when the sample loading step stops, the flow stops. Therefore, there will be some "purple" analyte still inside the sorbent bed. Some of the purple analyte resides in the interstitial spaces between the sorbent particles, the pores of the sorbent and filters and the cartridge outlet Luer type tip. The remaining compound of interest is eluted by adding an additional volume [at least the hold-up volume quantity for that cartridge] of the pure sample solvent, without any sample present, into the cartridge. See Figure 110. It will elute the remaining purple compound of interest [k is still 0] out of the cartridge, which is collected and then added to the volume that came through during the load step.

[SPE Strategy 1—Pass Through]

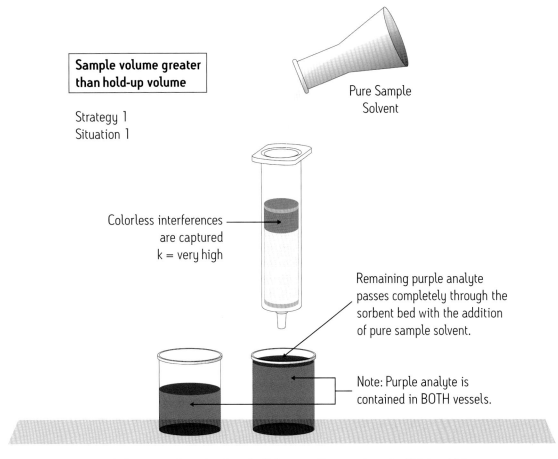

Figure 110: Pass-Through Strategy when the Sample Volume is Greater than the Hold-up Volume

The entire amount [mass] of the compound of interest is now eluted from the cartridge. Remember, this pure sample solvent will not be strong enough to move the interfering substances so that they will remain captured on the sorbent bed. Since the cartridge now contains only the "colorless" interfering substances, it can be discarded, according to hazardous waste regulations. The compound of interest has been separated from all the interferences. Combine the contents of the two vessels for further analysis.

2. Sample Volume Less Than Hold-Up Volume

In the second situation, under the same chromatographic conditions, the volume of sample loaded is less than the hold-up volume of the cartridge. Therefore, none of the "purple" compound of interest could have exited from the cartridge, since there was not enough liquid sample volume to allow it to reach the outlet of the cartridge. See the difference in Figure 111. The liquid that flowed from the cartridge, while the sample was being loaded, was the equilibration solvent that had been used in the previous step. The sample volume replaces only some of the equilibration solvent. What comes out during the load step can be discarded because it will not contain any of the purple compound of interest, even though $k = 0$.

[SPE Strategy 1—Pass Through]

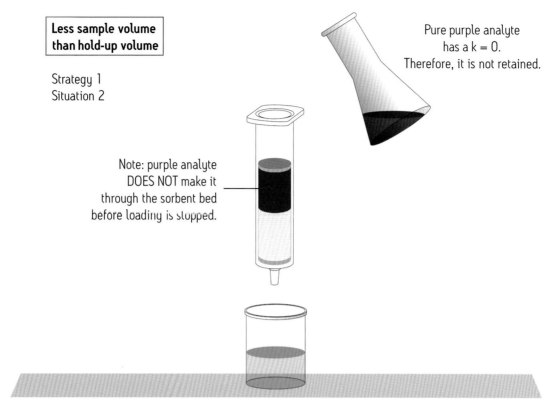

Figure 111: Pass-Through Strategy when the Sample Volume is Less Than the Hold-up Volume

In order to elute the compound of interest using the pass-through strategy, the next step is to pass approximately one to two times the cartridge hold-up volume of the pure sample solvent through the cartridge. This will elute the entire amount of the "purple" compound of interest [k = 0] which is collected for further analysis. Note that the purple analyte is only present in the second vessel. See Figure 112. The pure sample solvent will leave all of the "colorless" interfering substances retained on the sorbent bed, so that the SPE cartridge can now be discarded properly as before.

[SPE Strategy 1—Pass Through]

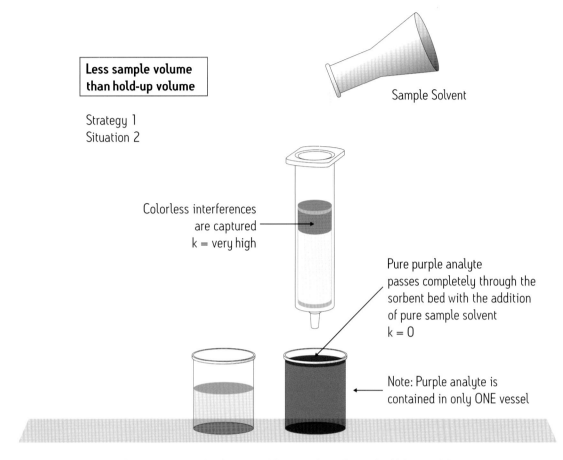

Figure 112: Pass-Through Strategy when the Sample Volume is Less Than the Hold-up Volume

One benefit of the pass-through strategy is that the SPE method has fewer steps, thus saving time and money. A typical protocol would include:

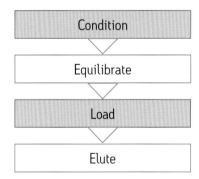

In summary, when there is more sample volume than the hold-up volume [Figure 110], the purple compound of interest is collected in both the load AND elute steps. When there is less sample volume, the purple compound of interest will be collected only in the elute step. See Figure 112.

[SPE Strategy 2—Capture]

Strategy 2: Capture the Compound[s] of Interest

This is the most common strategy for SPE, where a significant clean up of the sample matrix can be achieved by removing interfering substances that are not as strongly retained as the compound[s] of interest. The analytical chemist chooses the chromatography conditions that create a high attraction of the sorbent for the compounds of interest, and the compounds will be captured during the load step. To better see how this strategy works, we will use a new "sample" which appears green in Figure 113. This sample contains a less retained, yellow interference compound, a blue analyte of interest, and some additional colorless interferences that will be strongly attracted to the sorbent.

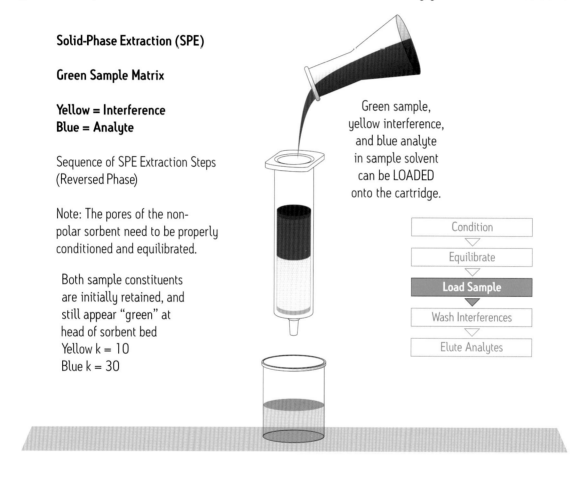

Figure 113: Capture Strategy—Load Step

As the sample matrix is loaded, all the compounds are well retained by the sorbent bed. The goal for Strategy 2 is to retain the blue analyte [high k = 30], while all the yellow interference, which has less attraction to the sorbent [lower k = 10], is washed [removed] from the cartridge. Note, that for any interferences with k = zero [no retention], some of the interference can leave the cartridge during the load step. This could occur if the sample volume being loaded is greater than the cartridge hold-up volume.

For the yellow interference in this experiment [k = 10], it would have less retention than the blue analyte [k= 30]. The yellow interference will be initially retained, but less strongly attracted to the sorbent than the blue analyte, or the more highly retained colorless interferences. A wash solvent will used that is strong enough to remove [release] the yellow interference from the cartridge, is used. However, this solvent should not be strong enough to release the blue compound of interest and the more highly retained interferences, which will remain captured. The liquids from the load and wash steps, which contain only the interfering substances, are discarded. Thus, we have reduced the number of remaining components of the sample matrix in the SPE cartridge. [Figure 114]

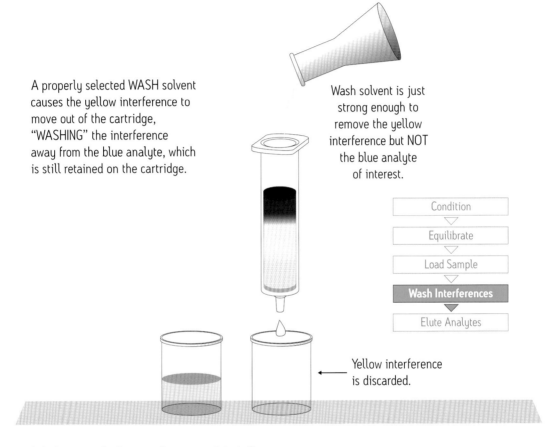

Figure 114: Strategy 2: Capture Strategy—Wash Step

What is still captured on the sorbent is the blue compound and any other colorless interferences that have the same or greater attraction to the sorbent as the blue compound. Two courses of action can be taken depending on the requirements of the subsequent analytical method.

First case, if the remaining colorless interferences will not create a serious analytical challenge, then all the remaining compounds, including the blue compound of interest, can be eluted with a sufficiently strong elution solvent and collected. The SPE cartridge is then disposed of properly. The subsequent analysis can then be performed.

Second case, if the colorless interferences, which still remain on the cartridge with the blue compound of interest, will cause problems in the subsequent analysis, then the elution step solvent is chosen differently. It must be strong enough to only elute the blue compound of interest, and leave the more attracted, colorless interferences retained on the sorbent bed. The elution step is collected, containing all the mass of the blue compound of interest. The SPE cartridge still contains the more retained colorless interferences and is disposed of properly. [Figure 115]

[SPE Strategy 2—Capture]

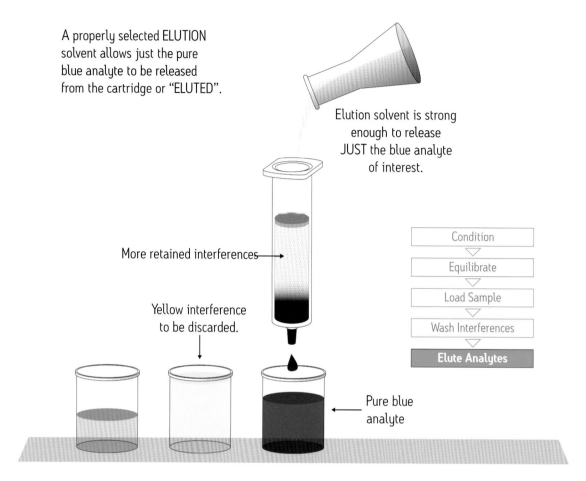

Figure 115: Capture Strategy—Elution Step, Second Case

To summarize, Strategy 2 is very powerful when the interfering substances in the sample matrix are less retained than the compounds of interest. Under the selected chromatographic conditions, the interfering substances can be removed from the compounds of interest by using properly chosen wash solvent[s]. The analyte is then released with a stronger elution solvent. A simplified SPE protocol includes:

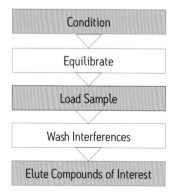

The compounds of interest are initially captured in the load step, while less-retained interfering substances are washed away. The elution solvent is selected, as described above, to release all the mass of the compound[s] of interest from the SPE cartridge.

[SPE Strategy 3—Capture and Fractionate]

Here the sample matrix contained interfering substances that have greater retention than the compound[s] of interest. This is handled by carefully selecting the elution solvent to be just strong enough to release the analytes, but NOT strong enough to release the interferences, which have a stronger attraction to the sorbent. They will remain in the cartridge and be discarded. For some methods, multiple wash and elute steps may be required to achieve the desired result.

Strategy 3: Capture and Fractionate the Compounds of Interest

This strategy is used when there are several compounds of interest that are best studied under different analytical conditions. It begins in the same way as the previous strategy, with the compounds of interest being captured in the load step. However, in this case, the specific goal is to separately elute the compounds in multiple elution steps using multiple, progressively stronger elution solvents. This is similar to an LC application using a step gradient. The process of using successive elution steps to separate the compounds of interest is called "fractionation." The eluate collected from each elution step contains different compounds that can then be further analyzed using optimal analytical techniques for that fraction.

An excellent example of this strategy is the complete analysis of all the compounds contained in the purple, grape flavored powdered beverage mix, Grape Kool-Aid® drink. This sample is a complex mixture of many components, including dyes, sugars, and flavorings. This approach can be useful in reverse engineering applications where it is important to identify and quantitate all of the components of a sample. Figure 116 shows the chromatographic power of SPE technology.

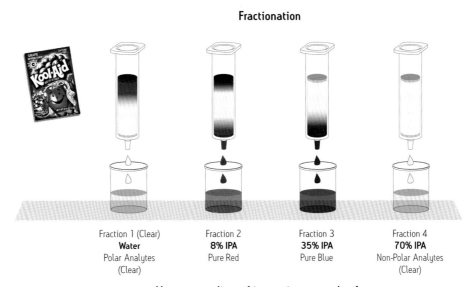

Figure 116: Capture and Fractionate Strategy

In this example, four different elution steps are performed, with each fraction containing different compounds of interest. The sample contained a mixture of very weakly retained clear polar compounds [acids and salts], a red dye, a blue dye, and very strongly retained non-polar flavoring oils, which also appear clear. The red and blue dyes are used to create the purple color of the original mixture.

[SPE Strategy 3—Capture and Fractionate]

All these compounds have different retention factors or different k values. Remember, the higher the k, the more strongly the compound is retained. By selecting the chromatographic conditions properly, compounds can be selectively separated from each other in SPE method. The dry powder was dissolved in water to create the sample matrix solution. A reversed-phase SPE device was chosen, which provides for a non-polar sorbent bed. For this method, all compounds which elute form the cartridge are further analyzed. Since none are discarded, there is no wash step. Following the conditioning and equilibration steps [not shown], the sample is loaded on to the cartridge. [Figure 117]

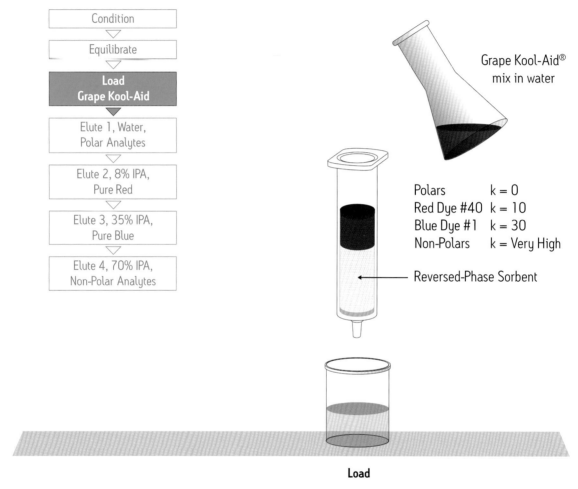

Not shown in Figure 117 are the first two steps of the protocol.

Figure 117: Capture and Fractionate Strategy—Load Step

Next it was determined that the first fraction [elute 1] would contain the polar sample compounds [acids and salts] that were not well retained. These appear "clear" when they elute from the cartridge first. See Figure 118. A relatively weak, elute 1 solvent [water, which is polar] was chosen to release these compounds. Still retained on the sorbent bed are the red and blue dyes as well as the other more strongly retained "clear" non-polar sample compounds. They all appear as the purple band at the top of the sorbent bed.

[SPE Strategy 3—Capture and Fractionate]

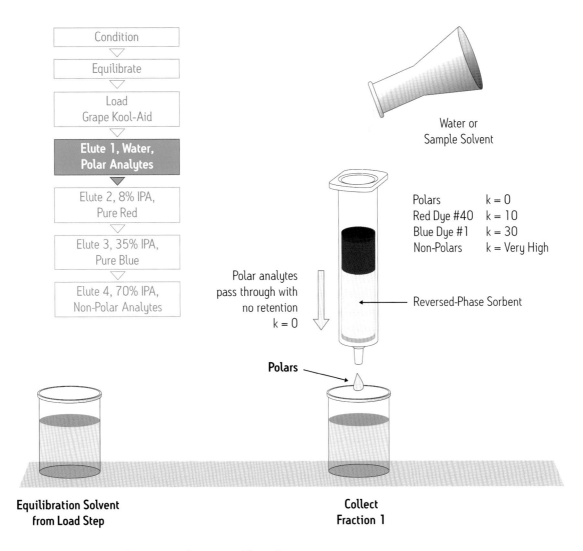

Figure 118: Capture and Fractionate Strategy—Elute 1

Note: How to properly select each solvent, and determine the required solvent volumes, will be discussed later. It is more important here to understand the concept of this strategy.

[SPE Strategy 3—Capture and Fractionate]

The second fraction was designed to contain only the red dye. A new, slightly stronger, elute 2 solvent [8% IPA in water] was selected to only release just the red dye away from the other more retained compounds. The red dye now has a k = 0, so that it is easily removed from the sorbent. See Figure 119. The red dye [elute 2 fraction] is collected and then analyzed. Note that the remaining compounds still retained by the sorbent bed are the blue dye and "colorless" compounds, which now appear as a blue band since the red dye, which had made the band appear as dark purple, has been eluted from the SPE device.

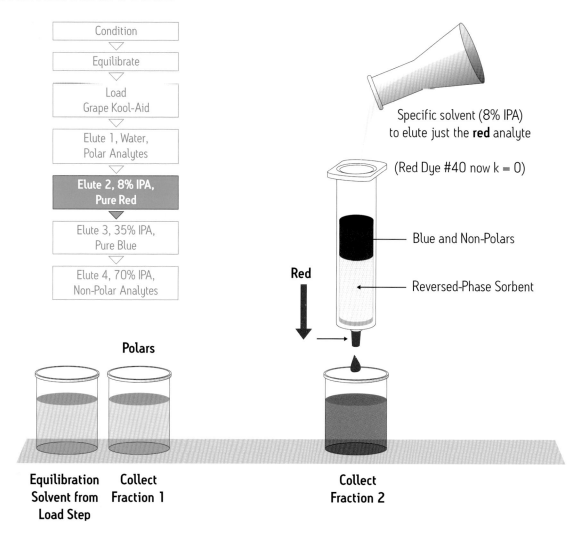

Figure 119: Capture and Fractionate Strategy—Elute 2

[SPE Strategy 3—Capture and Fractionate]

The third fraction was designed to contain only the blue dye. A new, even stronger elute 3 solvent [35% IPA in water] was selected to only release the blue dye while leaving all the other, more retained, "clear and colorless" non-polar flavoring oils captured by the sorbent. See Figure 120. The blue elute 3 fraction is collected and analyzed.

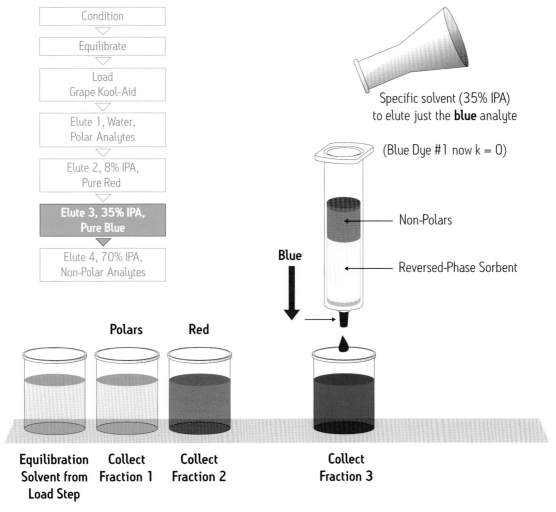

Figure 120: Capture and Fractionate Strategy—Elute 3

[SPE Strategy 3—Capture and Fractionate]

All the remaining sample compounds can now be released in the fourth fraction. A strong elute 4 solvent [70% IPA/water] is selected to remove all the remaining "clear and colorless" sample compounds. They are collected and further analyzed. The SPE cartridge is disposed of properly. [Figure 121]

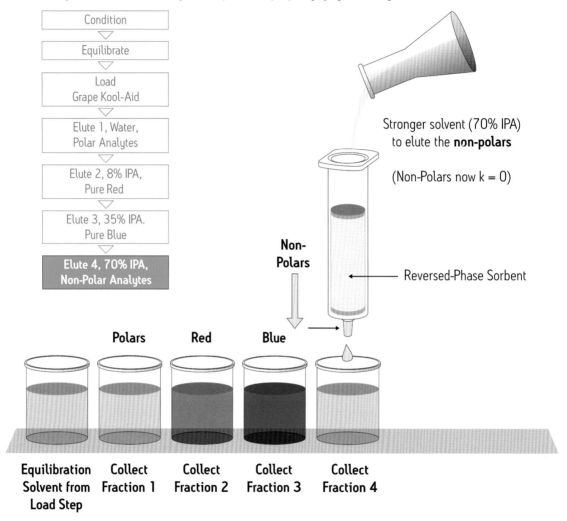

Figure 121: Capture and Fractionate Strategy—Elute 4

To summarize, Strategy 3 shows the powerful way an SPE cartridge can be used to selectively collect sample matrix compound classes, as well as collecting individual compound fractions. In SPE, the sorbent, combined with a step gradient of elution solvents of increasing strength, creates an extremely powerful sample preparation tool. The simplified SPE protocol for this sample would include:

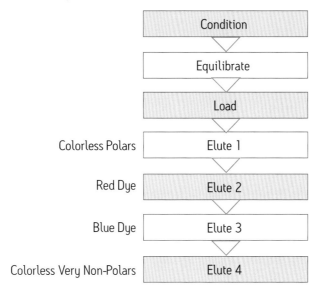

In this application, all of the sample matrix compounds were to be analyzed so no wash step was needed. The four fractions were created by a step gradient of progressively stronger elution solvents. Each fraction contained different compounds so that their subsequent analytical techniques could be optimized.

Strategy 4: Capture and Trace Enrich [Concentrate] the Compound[s] of Interest

In the world of sample preparation, this strategy has become incredibly important as the analytical sciences work to determine compound concentrations at lower and lower levels. Parts-per-billion [ppb], parts-per-trillion [ppt], and beyond are now required. Analytical instrumentation is becoming more and more challenged to reproducibly meet these goals. Instrument sensitivity may not be adequate to identify and quantitate these low level compounds in neat samples. Typical applications are in environmental, food safety, or metabolite and degradation analyses.

Trace Enriching Method

Strategy 4 provides the analytical scientist with a powerful approach to trace concentrate these compounds by continuously capturing them on the sorbent bed. Passing larger volumes [100 mL up to 1 L] of sample matrix through the SPE cartridge results in a buildup of a concentrated quantity of the trace compound[s] in a band at the inlet of the device. This occurs when the chromatographic conditions are created such that the compound[s] of interest has very a large k value, so that it is strongly captured by the sorbent bed. Figure 122 shows the initial loading of the sample.

[SPE Strategy 4—Capture and Trace Enrich]

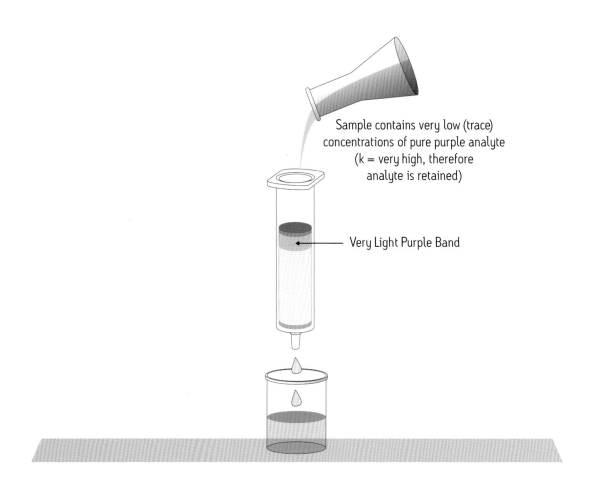

Figure 122: Using SPE for Trace Enrichment—Strategy 4

As more and more sample is loaded onto the cartridge this allows the concentration of the analyte to continue to increase. Note, it is important to carefully determine exactly how much sample volume was loaded onto the cartridge. This value will be used to back calculate the original concentration of the sample at the end of the analysis. See Figure 123.

[SPE Strategy 4—Capture and Trace Enrich]

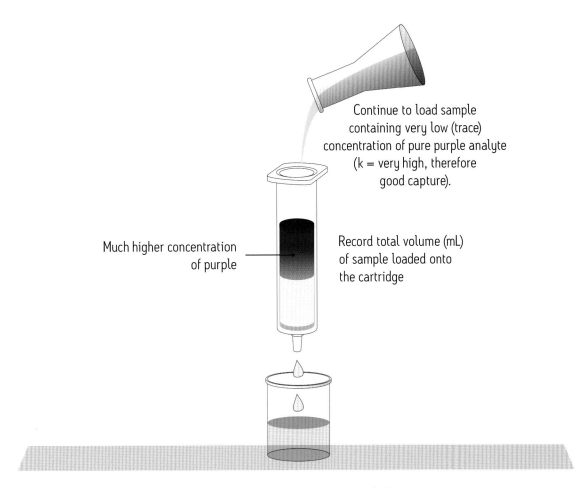

Figure 123: Trace Enrichment Strategy—Additional Sample Volume Loaded

The now concentrated compound[s] band, located on the inlet of the sorbent bed, can be eluted from the cartridge using a strong elution solvent to release it. See Figure 124. Note: in some applications you may use a wash step to flush away any interference before eluting the compounds of interest.

[SPE Strategy 4—Capture and Trace Enrich]

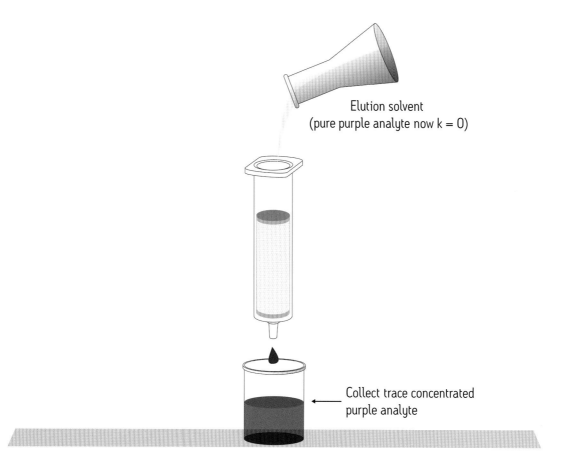

Figure 124: Trace Enrichment Strategy—Elution Step

This new concentrated sample can then be analyzed and the amount of analyte determined. The calculation of the original analyte sample concentration is shown on page 137.

Notice that this elution step was performed in the normal flow direction. Since the purple analyte band must now be flowed through the full sorbent bed for complete elution, a certain volume of elution solvent will be required. However, if we reverse the flow direction, as shown in Figure 125, less elution solvent is needed since the analyte only has to move off the original inlet of the cartridge sorbent bed.

[SPE Strategy 4—Capture and Trace Enrich]

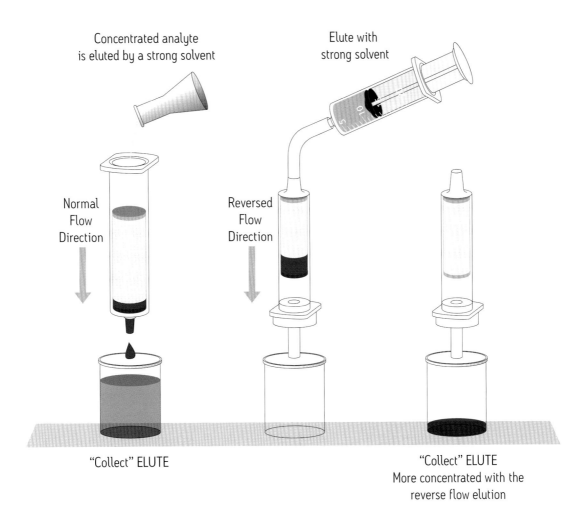

Figure 125: Trace Enrichment Strategy—Reversed Flow Elution Step—Syringe Design

By configuring the SPE device using available adapters for the syringe design, or just switching the ends of the Plus design, a reversed flow direction can be achieved [Figure 126]. The compound band can be eluted from the inlet of the cartridge, with less elution solvent which means a much higher concentration of the analyte.

[SPE Strategy 4—Capture and Trace Enrich]

Figure 126: Trace Enrichment Strategy—Reversed Flow Elution Step—Plus Design

The simplified SPE protocol for this strategy would include:

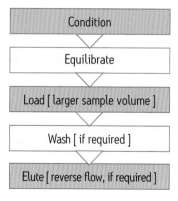

[SPE Strategy 4—Capture and Trace Enrich]

Calculating the Results

The trace enrichment of compounds by SPE provides a final trace concentrated sample that can be easily quantified by the analytical technique since the concentration levels are now in the sensitivity range for the detector. The analyte concentration in the original sample can then be accurately determined by performing the following calculation:

$$\text{Original Sample Concentration} = \frac{\text{Analytical Result [mass] of Concentrated Sample}}{\text{Volume of Original Sample loaded on SPE}}$$

The analyte concentration in the original sample is calculated by taking the mass from the analytical result of the concentrated sample and dividing by the volume of original sample that had been processed through the cartridge. SPE Strategy 4 provides the ability to greatly increase the sensitivity of the overall analytical determination.

This discussion on common SPE strategies has shown both the power and versatility of this sample preparation technique. At the beginning of the method development process, consider the four different strategies and how they may be best used to solve your problems.

Now that the various strategies have been explained, the next step is to decide which strategy to use, and then pick the chromatographic mode and sorbent technology to make it work.

[Chromatographic Mode Selection]

Chromatographic Mode Selection

Table 9 summarizes the various modes of SPE, and is a good starting point for the method development process.

Table 9: Chromatographic Mode Selection with Sorbents

	Reversed Phase	Normal Phase	Ion Exchange	
Analyte	Moderate to low polarity	Low to high polarity/ neutral	Charged and ionizable	
Separation Mechanism	Separation based on hydrophobicity	Separation based on polarity	Separation based on charge	
Sample Matrix	Aqueous	Non-polar organic solvent	Aqueous/low ionic strength	
Condition/Equilibrate SPE Sorbent	1. Solvate with polar organic 2. Water	Non-polar organic	Low ionic strength buffer	
Preliminary Wash Step	Aqueous/buffer	Non-polar organic	Low ionic strength buffer	
Elution Steps	Increase polar organic content	Increase eluotropic strength of organic solvent mixture	Stronger buffers—ionic strength or pH to neutralize the charge	

			Anion Exchange	Cation Exchange
Sorbent Functionality	C_{18}, tC_{18}, C_8, tC_2, CN, NH_2, HLB, RDX, Rxn, RP	Silica, Alumina, Florisil, Diol, CN, NH_2	Accell™ Plus QMA, NH_2, SAX, MAX, WAX	Accell Plus CM, SCX, MCX, WCX, Rxn, CX
Sorbent Surface Polarity	Low to medium	High to medium	High	High
Typical Solvent Polarity Range	High to medium	Low to medium	High	High
Typical Sample Loading Solvent	Water, low strength buffer	Hexane, cholorform, methylene chloride	Water, low strength buffer	Water, low strength buffer
Typical Elution Solvent	MeOH/water, CH_3CN/water	Ethyl acetate, acetone, CH_3CN	Buffers, salts with high ionic strength, increase pH	Buffers, salts with high ionic strength, decrease pH
Sample Elution Order	Most polar sample components first	Least polar sample components first	Most weakly ionized sample component first	Most weakly ionized sample component first
Mobile Phase Solvent Change Required to Elute Compounds	Decrease solvent polarity	Increase solvent polarity	Increase ionic strength or increase pH	Increase ionic strength or decrease pH

Choosing the Right Device Size

The next two topics deal with how best to determine what size SPE device to use. Our goal is to help you choose the smallest size cartridge sufficient for the goals and conditions of your method. This will save you money on cartridges, solvents, and overall time to complete your method. Choosing the smallest cartridge size can also provide enhanced sensitivity by keeping concentrations at higher levels. All the different type and size cartridges can be found on the Waters website.

One of the most common questions scientists ask is, "How much can I load onto this cartridge?" Unfortunately, there is no easy answer. It depends on the conditions in your method and the size of the sample you need to prepare. Two studies are highly recommended during the method development process: mass balance and breakthrough. When properly done, these studies will help improve the robustness of your new method and reduce the cost per analysis. Please refer to the "Key Terms and Calculations" section for the proper way to calculate % recovery, and, if you are using a mass spectrometer, the calculation for matrix effects.

Note: These topics can appear to be very foreign or confusing when first working with SPE methods development. The following section is designed to develop the concepts using diagrams and explanations which will help you understand why the science works, and how valuable these studies are in creating powerful and robust methods.

Mass Balance in Method Development and Troubleshooting

Understanding the Concept

When performing SPE, understanding what is happening to the compound[s] or analyte[s] of interest is important to ensure dependable sample preparation performance. We must understand where the compound[s] of interest is, as the protocol steps are performed. In some cases, what we would expect to happen is not what physically has occurred. This causes poor analytical results.

The concept of mass balance provides assurance that the location of the compound[s] of interest has been experimentally determined, by quantifying where the mass of each compound is located, as the steps of the protocol are performed. This mass balance determination is done as a part of a proper methods development process to ensure robust performance. It can also be used in troubleshooting an existing method.

Remember that the location of a compound will be determined by the chromatographic conditions in the SPE cartridge. A weakly retained compound will flow relatively quickly through the sorbent bed and be completely released from the cartridge with a minimal volume of solvent. However, a strongly retained compound will move very slowly through the cartridge and require a significant volume of solvent to release it completely. That is why a mass balance determination is important to ensure the performance that is expected.

[Mass Balance]

Some Practical Examples—Where is my Analyte?

The following diagrams illustrate the important concept of mass balance in SPE.

On the left in Figure 127 is an SPE cartridge, with a hold-up volume [cartridge volume] of 2 mL. [See "Key Terms and Calculations" section for more information on the hold-up volume on page 62]. The purple sample contains a solvent solution of two dyes, a red dye; with a k = 1 and a blue dye with a k = 30 under the loading conditions. In a chromatographic system, the value of k indicates the volume of that solvent that will be required for the compound to completely pass through the sorbent bed. Note that the red dye begins to separate but does not reach the outlet of the cartridge when the 2 mL load stops. This is because there is some slight retention of the band [k = 1] but not near as much as the retention for the blue dye, which remains captured at the head of the cartridge.

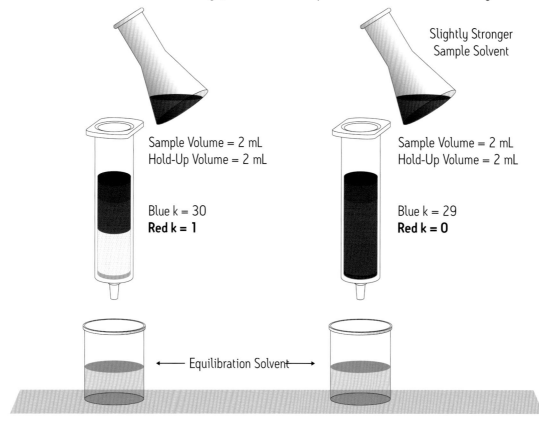

Figure 127: Mass Balance—Impact of k Value

Now let's make the sample solvent a little stronger and follow what happens in the diagram on the right in Figure 127. Here the red dye is k = 0, and the blue dye is k = 29. Note how some of the red dye proceeds all the way to the end of the cartridge. This occurs because k = 0, and the cartridge volume and the sample volume are both the same. However, no red dye elutes yet.

For complete elution of the red compound using the original sample solvent, the following simple calculation applies:

$$\text{\# Cartridge volumes [CV] of solvent} = [k \text{ value}] \times CV + 1\ CV$$

[Mass Balance]

In Figure 128, the cartridge volume = 2 mL, sample volume = 2 mL, and k = 1 for the red compound. See diagram on the left. Therefore, the amount of elution solvent required would be two cartridge volumes [4 mL] to completely elute that compound from the cartridge using the solvent [see diagram on left]. If a different, stronger solvent was used, which created a k = 0 for the red compound, then only 2 mL of this new solvent would be required [see diagram on right]. For example, if a compound with a k = 0 is loaded onto the SPE cartridge, it will take one additional cartridge volume of the same solvent that is in the sample to completely elute that compound from the cartridge. Note: To be conservative, add an additional $1/2$ CV of elution solvent volume to ensure complete elution of the compound.

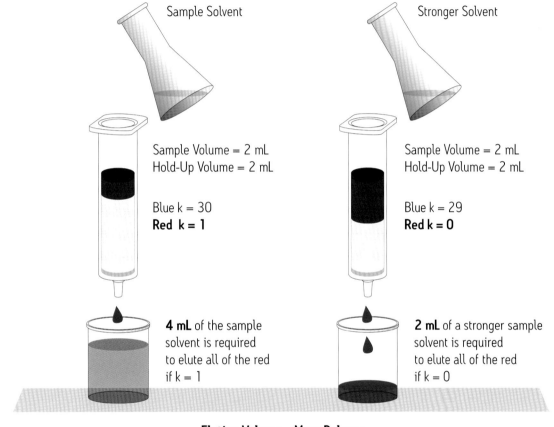

Figure 128: Mass Balance—Changing the k Value

[Mass Balance]

Figure 129 shows what happens when 1 mL of the purple sample [on the right] is loaded, compared to 2 mL [on the left] as before. Note that the red dye band has moved about one sixth of the way down the sorbent bed, since the hold-up volume is 2 mL and k = 1 for the red dye. The blue dye remains in a tight band at the head of the sorbent bed because it has a much higher k = 30, and does not move significantly down the sorbent bed. That portion of the cartridge appears purple because there is some red dye still mixed in with the blue dye when the flow stops at the end of the loading step.

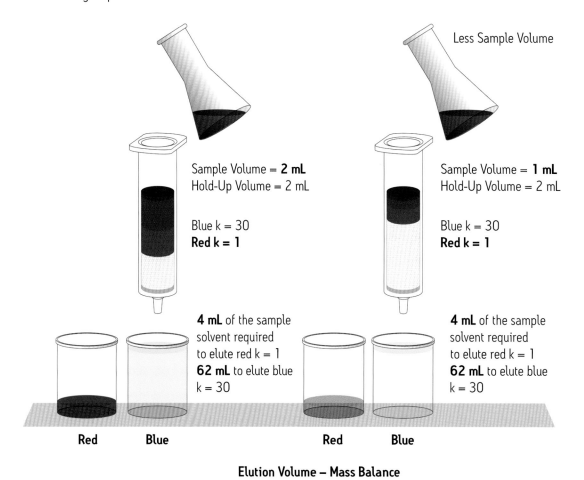

Figure 129: Mass Balance—Changing Sample Volume

Keeping the same solvent that was used in the sample, the elution volumes remain the same. It would require ~4 mL of the sample solvent to completely elute the red dye. Interestingly, to elute all of the blue dye with that same solvent, k = 30, it would require 31 cartridge volumes of solvent, which equals 62 mL! Note, the concentration values would be one half for the 1 mL sample, since one half of the sample was loaded.

To completely elute the blue dye with k = 30, it would take 62 mL of weak sample solvent! That is why we use stronger solvents in wash and elution steps to drop the k values substantially, thus requiring far less of the stronger solvent.

[Mass Balance]

In SPE, strong solvents [step gradients] are used to elute compounds in much less volume, as seen in Figure 130. A new, much stronger solvent, which now has a much higher attraction for the blue dye [higher than the sorbent] will reduce the k of the blue dye to 0. Therefore, to completely elute the blue dye with this new solvent would require only ~2 mL. This yields a much more concentrated fraction, which greatly enhances the sensitivity of the SPE protocol.

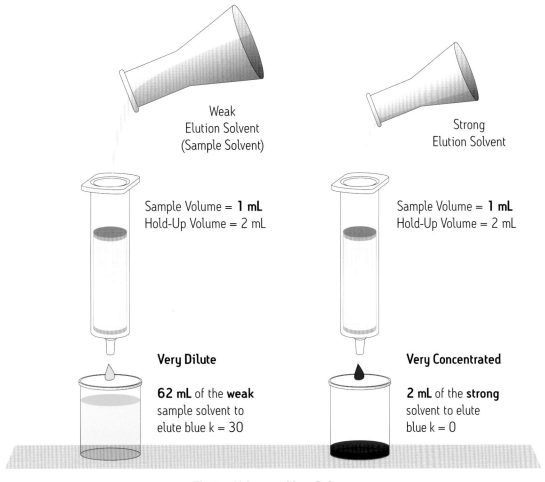

Figure 130: Benefits of Strong Elution Solvents

[Mass Balance]

Choosing Your Strategy

It is important to keep in mind the behavior of compounds in a chromatographic system [Figure 130] when performing the mass balance study. Remember, this study is only necessary during method development, in order to ensure robust performance.

To make a preliminary mass balance determination for a planned SPE protocol, first select one of the four strategies and then select the sorbent type and size [sorbent mass] of SPE cartridge that will be used.

All four strategies share the same eight common steps, described below from A to H. Any additional steps will depend on which strategy you choose.

Common Steps for All Strategies:

A. Determine the hold-up volume as described previously. [See page 62]

B. Obtain standards for the compound[s] of interest and add them to the same sample solvent as the actual sample. Make up the standards in the sample solvent to approximate the concentration of the compound[s] of interest that will be in the actual sample.

C. Measure out the same volume of the prepared standards test sample that will be used with the actual sample later in the final protocol. Calculate the mass [concentration x volume] of each of the standards that will be loaded onto the SPE cartridge.

D. Compare this volume to the hold-up volume to help predict where the least retained compounds will be after loading.

E. Initially select a weak wash solvent [and volume] and a strong elution solvent [and volume] depending on the protocol strategy that is to be used.

F. Prepare the SPE cartridge following the conditioning and equilibration steps.

G. Set up vessels to collect everything that exits the SPE cartridge.

H. Load the specified volume of the prepared standards test sample, collect what exits the cartridge, and label it "Collect Load."

Strategy 1: Pass Through

We are going to use a green sample made up of a yellow standard with a k = 0 as our compound of interest, and a blue standard with a k = 30 as our interference. For the first experiment, the sample load volume is greater than the hold-up volume of the cartridge. [Figure 131]

[Mass Balance]

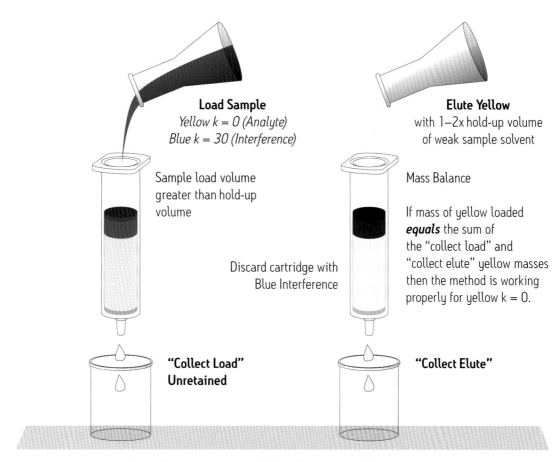

Pass Through Strategy – 1 Mass Balance for Yellow Analyte
Sample Volume GREATER than Hold-Up Volume

Figure 131: Mass Balance—Pass-Through Strategy—Complete Elution Step

Here, we must be concerned that the un-retained yellow standard may have eluted from the cartridge during the load step [diagram on the left]. This will occur if the sample load volume was greater than the hold-up volume. Some will pass through and be present in the "collect load." However, due to the hold-up volume after the load step stops, some of the un-retained yellow standard will still be in the cartridge. To remove it completely [diagram on the right], an elute step is performed using pure sample solvent with a volume of ~1–2x of the hold-up volume, as described in Step I below.

Additional Steps for Strategy 1 [See page 144 for steps A–H]

Complete the mass balance determination by performing the following steps.

I.   Elute with ~1–2x the hold-up volume of pure sample solvent [no standards]. Collect what exits from the cartridge, and label it "Collect Elute."

J.   Analyze the "Collect Load" and the "Collect Elute" separately. Add the two mass values for the yellow standard together, which should total the original mass that was initially loaded on the SPE cartridge for that standard in Step H [see Figure 131 on the right]. This is a complete mass balance, since the mass of the yellow standard can be followed through the steps of the protocol, and all of the mass of the standard has been accounted for.

[Mass Balance—Pass Through]

However, if the total mass for a standard does not equal the amount originally loaded onto the cartridge [see Figure 132 on the right], then this protocol will have to be modified. The mass balance below indicates that some of the standard was still left on the sorbent bed, because it must have had a slight amount of retention, $k \neq 0$.

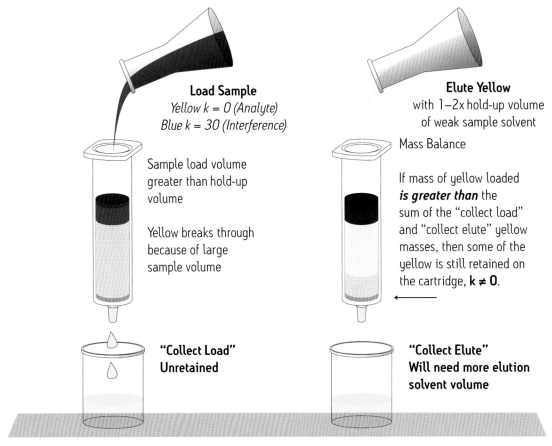

Pass Through Strategy –1 Mass Balance for Yellow Analyte
Sample Volume GREATER than Hold-Up Volume

Figure 132: Mass Balance—Pass-Through Strategy—Incomplete Elution Steps

This condition requires a larger volume of original elute solvent, or a change to a somewhat stronger elute solvent. Performing a mass balance during the method development process provides important information to ensure robust performance.

For Strategy 1, if the test standard sample load volume was less than the hold-up volume, then all of the standard must be in the "Collect Elute" vessel only, since none of the yellow compound could have made it all the way through the cartridge during the loading step. Check that the mass of yellow in the load step is equal to the mass in the elution step. Having the sample load volume less than the hold-up volume provides for a simpler method for Strategy 1.

Strategy 2: Capture

In Strategy 2, compounds of interest will initially be retained by the sorbent bed. A wash step is used to remove interferences that were not as strongly retained as the compounds of interest. Using the same green sample, now the yellow standard is now will be the interference substance and the blue standard is the compound of interest. The sample volume is greater than the hold-up volume. [Figure 133]

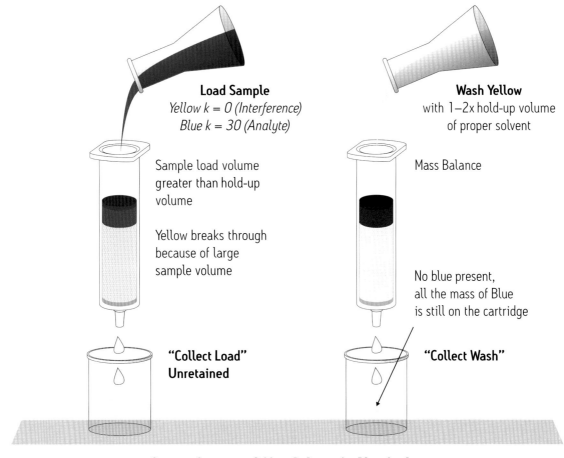

Capture Strategy –2 Mass Balance for Blue Analyte
Sample Volume GREATER than Hold-Up Volume

Figure 133: Mass Balance—Capture

[Mass Balance—Capture]

Additional Steps for Strategy 2 [See page 144 for steps A–H]

Complete the mass balance determination by performing the following steps:

I. Perform the wash step with the specific volume of solvent desired. Collect what exits the cartridge and label it "Collect Wash." The yellow standard should be completely removed from the cartridge and the blue standard should still be retained by the sorbent bed. [Figure 133]

J. Analyze the "Collect Wash" and make sure that no quantity of the blue standard has been removed from the cartridge. Note: Sample matrix interferences may cause some problems in the LC analysis for the blue standard. If an LC analysis is not possible, then complete steps K and L and determine if the % recovery is low. It could be that some of the blue was washed away. Adjust as necessary.

If some of the blue standard is present in the "Collect Wash," then the mass balance indicates that the wash solvent and or the volume of wash must be modified. Use less of this wash solvent or change to a weaker wash solvent.

K. Perform the elute step which is designed to release all of the blue standard, using the specific volume of the strong elute solvent desired. Label what exits the cartridge as "Collect Elute." [Figure 134]

L. Analyze the "Collect Elute" vessel and determine if all of the mass of the blue standard has been released from the cartridge. This should equal what was loaded in Step H. If it is fully released, then the mass balance indicates that the preliminary protocol is working properly.

[Mass Balance—Capture]

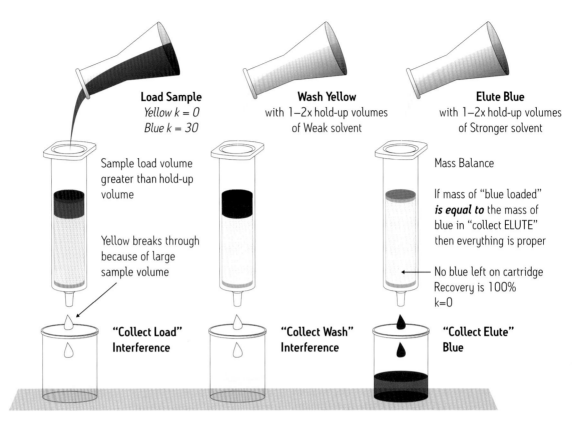

Capture Strategy −2 Mass Balance for Blue Analyte
Sample Volume GREATER than Hold-Up Volume

Figure 134: Mass Balance—Capture—Complete Elution Step

If the mass of the blue standard does not equal its load mass, then the mass balance indicates that some of it is still inside the cartridge or some of the blue standard was washed away in Step I. A modification to the protocol will be required. Select an even stronger elute solvent with the same volume, or increase the volume of the original elution solvent.

Strategy 3: Capture and Fractionate

Strategy 3 features multiple fractions being collected for specialized analysis. Compounds will be retained in a range from weak to strong. Typically, multiple elution steps are performed using different strength solvents in a step gradient. A mass balance determination is critical to ensure that the expected compounds are present in their designated fractions. For this discussion, we will use a more complex sample with compounds that have a wide range of polarities, resulting in different retention factors, or k values. Sample volume is greater than hold-up volume. [Figure 135]

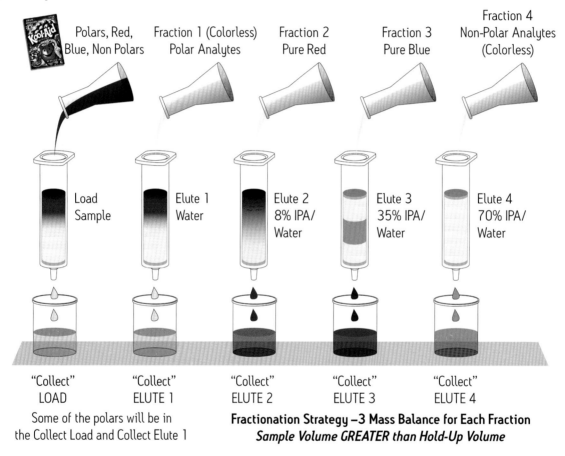

Figure 135: Mass Balance—Capture and Fractionate

Additional Steps for Strategy 3 [See page 144 for steps A–H]

Complete the mass balance determination by performing the following steps:

I. After the load step, label what exits the cartridge as "Collect Load." Depending on the hold-up volume, this could contain some of the very weakly retained standards. In this experiment, the load volume was greater than the hold-up volume so some of the early eluting very polar compounds passed all the way through the cartridge and were found in "Collect Load."

The elute 1 step is typically made using the pure sample solvent. The elute 1 volume will be ~1–2x the original sample load volume. Label what exits the cartridge as "Collect Elute 1." Fraction 1 will contain the weakly retained compounds; however, in many cases, fraction 1 is created by combining "Collect Load" and "Collect Elute 1."

J. Analyze "Collect Load" and "Collect Elute 1" vessels separately and then combine the numerical results for each of the weakly retained standards. Determine if all of the standards that were supposed to be in Fraction 1 were released and collected. This is done by comparing the total mass results for each standard with the mass of those compounds from the load step.

If any of the weakly retained standards that are supposed to be in fraction 1 are not fully released, then strengthen the elute 1 sample solvent and/or increase the volume of the solvent.

If any of the standards that are supposed to be in fraction 2 have eluted too early in fraction 1, it is possible that the elute 1 sample solvent volume was too high. Adjust as necessary.

K. Continue with the elute and analysis steps for the additional fractions. The mass balance determination ensures that each fraction contains only the desired compounds. Adjust the elution volumes or solvent strength as needed. Once all the masses for each of the compounds of interest are accounted for, this will ensure that the protocol will be robust.

Strategy 4: Capture and Trace Enrichment

Strategy 4 is the trace enrichment or trace concentration strategy. As described on page 131, the compound[s] of interest is present in very low concentration. The chromatographic conditions are selected to strongly capture these compounds on the sorbent bed by creating a very high retention [high k value] for them.

A large volume of sample is loaded onto the SPE cartridge. The sample volume must be carefully measured and recorded, because it will be used in the final compound[s] concentration calculation. Most of the interfering substances contained in the sample matrix will have little retention and will exit from the cartridge during the loading step. In this experiment, purple represents compounds of interest at very low levels. As the enrichment continues, these compounds will have much higher concentrations. [Figure 136]

[Mass Balance—Capture and Trace Enrichment]

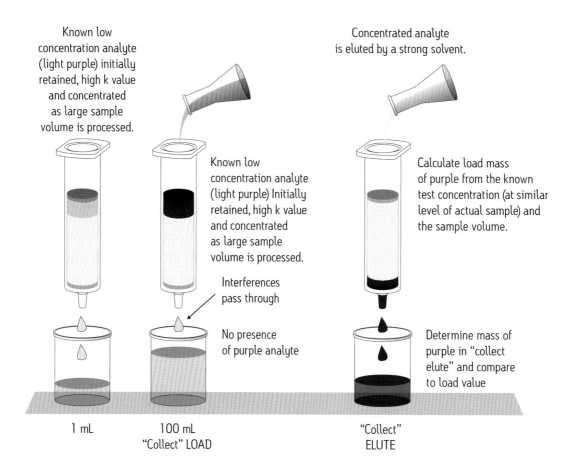

Figure 136: Mass Balance—Trace Enrichment

Additional Steps for Strategy 4 [See page 144 for steps A–H]

In trace concentration methods, you will not always know the starting concentration of your compounds. You will have to ensure that the k values are very high during the load step and account for any variation in the sample matrix that could bring in interferences. The interferences will take up sites on the sorbent, thereby reducing the amount of compound of interest that can be captured. Also, the sample volume should always be specified. This study will ensure that you do not overload the sorbent bed and lose your compounds of interest. Testing the "Collect Load" for any presence of the compounds of interest provides assurance that there is complete capture. Challenge the SPE device with spiked samples with a range of concentrations and typical variations in the sample matrix.

For the elution step, determine which solvent and what volume is needed to completely elute the compounds of interest.

Complete the mass balance determination by performing the following steps:

I. After the load step, label what exits the cartridge as "Collect Load." Analyze this for any evidence of the compound[s] of interest. None should be detected. If they are detected, then the mass of the sorbent bed should be increased [a longer sorbent bed cartridge] or a different type of sorbent should be selected. The longer sorbent bed cartridge will provide additional capacity for the compound[s] because they would have to travel farther before exiting the cartridge. In some cases reducing the sample volume loaded onto the cartridge could be considered as long as it does not impact the overall accuracy of the analysis.

J. In some cases, a wash step with a slightly stronger solvent may be needed if there are any moderately retained interferences that would affect the analysis. Label what exits as "Collect Wash" and test that none of the compounds of interest are present. If they are, then the wash solvent, or its volume, must be changed.

K. Elute the compound[s] of interest using a stronger elution solvent to release it from the sorbent. Label what exits from the cartridge as "Collect Elute."

L. Analyze the "Collect Elute" vessel for the compound[s] of interest. All of the mass of the compound[s] should be eluted from the cartridge. If there are any highly retained interferences along with the compounds of interest that could cause problems in the analysis, then reduce the strength or volume of the elution solvent and test for complete recovery of the compounds of interest.

If complete recovery of compound[s] was not obtained, the mass balance indicates that some of the compound mass is still inside the cartridge. Several adjustments can be made such as selecting a stronger or larger volume of elute solvent. When you try a stronger solvent, always make sure that no problematic interferences are eluted with your compounds of interest.

Reverse Flow for Higher Concentration

In Figure 136, the elution step flow was in the normal direction. In order to get a concentrated band of the compounds of interest completely out of the sorbent bed, a volume of elution solvent was used. As the band proceeded down the length of the sorbent bed it became somewhat more dilute. For samples where maximum enrichment is critical, there is an elution step technique that solves this problem. The elution step is carried out in the reverse flow direction which provides a significant improvement in the final concentration of the analytes. [Figure 137]

Adapters are available to configure the various cartridge designs to achieve flow in the reverse direction.

[Load Capacity/Breakthrough]

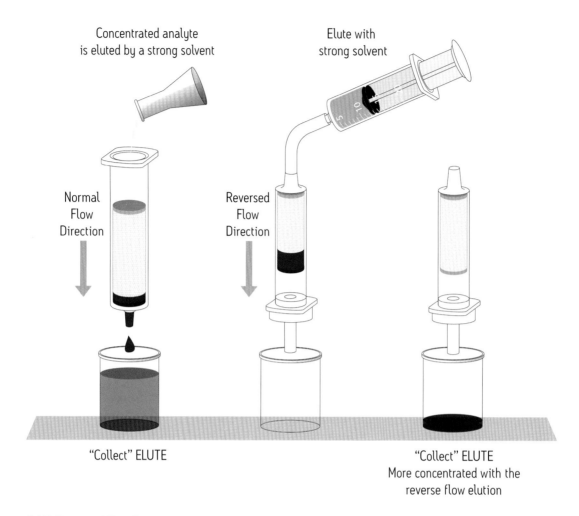

Figure 137: Reversed Flow Direction for Elution Step

Reverse flow direction allows for more complete release of the trace concentrated compound[s] captured at the inlet of the sorbent bed, with the added benefit of needing less elution solvent volume since the compound[s] do not have to travel through the full length of the sorbent bed. This study is performed after you have your SPE method steps initially developed and before you perform the Breakthrough Study.

A mass balance determination provides the analytical chemist with the precise information needed for the development of a robust SPE method. This helps avoid future problems when running the method.

Load Capacity/Breakthrough

Probably the most common question in SPE is "How much sample can I load onto this cartridge?" Although it sounds like a very simple request, no correct answer can be given unless the chromatographic conditions are defined for that particular situation. For each application, there will be a specific answer for a specific cartridge, and there is only one dependable way to determine that answer: A Breakthrough Study.

[Load Capacity/Breakthrough]

A breakthrough study determines the capacity for a SPE cartridge, or, a point at which the cartridge will not have sufficient capacity to retain all of a compound of interest in the sample volume that is loaded. When this capacity limit is reached, any additional compound being loaded will "breakthrough" the cartridge unretained, resulting in poor SPE recoveries. This study is performed after you have your SPE method steps initially developed and after you do the mass balance study.

The breakthrough study will help determine the best choice in SPE cartridge type and size. In general, choose the smallest size cartridge possible while still meeting your method development goals. This will reduce costs for cartridges, and will also reduce solvent volumes, which increases compound concentrations, improves sensitivity, reduces the processing time for the method, all contributing to lowering costs.

Importance of Chromatographic Conditions

The reason that a breakthrough study is required relates to all the factors that contribute to the capacity for a given SPE cartridge. It is obvious that the size of the cartridge is important, specifically the mass of sorbent that it contains [i.e. 100 mg vs. 500 mg], as well as the type of sorbent [i.e. C_{18} vs. C_8 vs. CN]. However, far more important are the chromatographic conditions that exist inside the SPE cartridge while the protocol is being performed, especially during the loading step. The chromatographic conditions are affected by the following factors:

- Sorbent type
- Sorbent mass
- Sample solvent
- Sample matrix substances [and any variations]
- Sample matrix/analyte concentrations
- Sample volume
- Compounds of interest
- Retention factor, k, for the compounds of interest "charge state for ionizable compounds"
- Retention factor, k, for interfering substances*
- Flow rate of the load step

A properly performed breakthrough study will provide an accurate answer to the capacity question for that application, resulting in a more robust method. It may seem like extra work, but the benefit to the methods development scientist will be a much more robust protocol that should provide excellent results over time. Without this study, variations in % recovery values can result due to overloading of SPE devices that do not have sufficient capacity.

The study is completed during the methods development process, after the overall SPE protocol strategy, sorbent type, and wash and elute solvents have been chosen. Based on the sample volume to be loaded, an initial cartridge size is selected for sorbent mass. In general, larger volumes and more concentrated sample matrices will require a greater mass of sorbent. The breakthrough study uses the actual sample matrix with the analytes spiked to known levels. This allows for the proper determination of the SPE cartridge capacity with all the sample matrix compounds present.

*More information on the Retention Factor, and k, can be found in the Glossary on page 187.

[Loading the Sample]

Breakthrough Study Experiment

This study will consist of a series of experiments where increasing sample matrix volumes with the analytes of interest at known concentrations will be loaded onto several cartridges of the same size. The recovery results for the analytes are then compared. The steps are outlined below and diagrams will be used to make it easier to understand.

Take 5 to 6 SPE cartridges of the same size and sorbent type. Condition and equilibrate the cartridges.

Label each cartridge with the sample volume to be loaded. Bracket the intended volume with 1 or 2 values lower in volume and 1 or 2 higher in volume [i.e. if the intended sample matrix volume is 1 mL, set the others at 0.5, 0.75, 2.0, and 3.0 mL].

Loading the Samples

Load the specified sample volume for each cartridge at the same flow rate. Record the mass of each compound of interest that was loaded based on its concentration and sample volume. See Figure 138. Here we have 6 cartridges* each loaded with a different volume of sample, ranging from 0.5 to 5 mL. The sample is a mixture of "yellow" and "blue" compounds, with the yellow compound having a lower value of $k = 0$, resulting in less retention relative to the blue compound, $k = 2$.

* Intended Sample Volume = 2 mL and Hold-Up volume ~ 1.1 mL

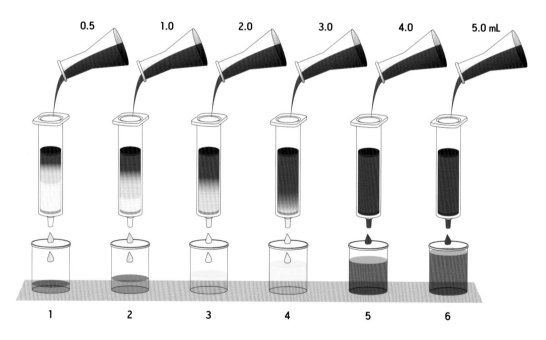

Figure 138: Breakthrough Study—Loading

Follow the remaining steps of the SPE protocol and collect the elution step in a vessel labeled with the corresponding sample volume.

Analyze each vessel to determine the mass recovered for each compound of interest. Calculate the % recovery for each compound and each sample matrix volume based on the amount loaded.

Plot the % recovery for each compound versus the sample matrix volume. When the % recovery for a compound falls below 100%, then breakthrough has occurred for that compound at that sample volume. See Figure 139.

Based on the plot, determine the maximum sample matrix volume for this method.

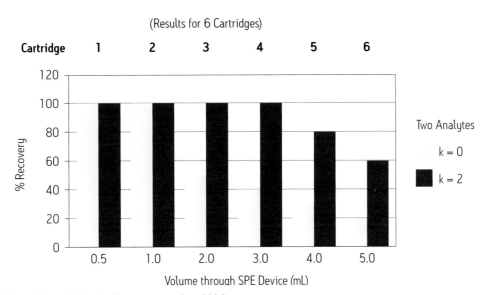

Figure 139: Breakthough Study: Recovery vs. Load Volume

Let's follow this example more closely. The sample includes two compounds of interest, a yellow dye and a blue dye. The sample therefore appears "green." In real life, your analytes would typically be clear and colorless. You would use the response from your detection system to calculate where your analytes of interest are located during the steps of the protocol.

The concentration of the sample matrix and analytes is held constant for all the experiments. Under the chromatographic conditions present in the SPE cartridge, the yellow dye has a $k = 0$, and the blue dye has a $k = 2$. This indicates that the blue dye is more retained by the sorbent than the yellow dye. Remember, the lower the k value the less the analyte is retained by the sorbent. Our intended sample volume is 2 mL, and an initial SPE cartridge size was also selected. For the study, six sample load volumes were evaluated, [0.5, 1.0, 2.0 (intended), 3.0, 4.0 and 5.0 mL]. We perform the loading step as described.

The "green" sample was loaded onto each of the cartridges with six different specified volumes. Viewing from left to right, Figure 138 the first cartridge had 0.5 mL loaded, the next cartridge had 1.0 mL loaded, and the remaining cartridges were as specified. It is interesting to follow the location of the two dyes at the end of the load step. Under the chromatographic conditions in the cartridge, which are determined by the sorbent type, sample solvent, and the analyte, the yellow dye has a $k = 0$, and the blue dye compound has a $k = 2$.

[Loading the Sample]

0.5 mL Load Volume

In Figure 138, when 0.5 mL of sample being loaded, there is a slight separation of the two dyes. As expected, the yellow dye is moving faster, because it has a lower k, meaning less retention by the sorbent bed. Some of the yellow has separated away from the inlet of the sorbent bed. The head of the bed still appears green because the "blue" band is not moving; however, there is some "yellow" still present in that "blue" region when the flow is stopped at 0.5 mL. Therefore, that band appears "green." Note that the sorbent bed has additional capacity, since neither of the dyes has reached the outlet of the cartridge during this loading step. The liquid that exits the SPE cartridge is collected and labeled.

We will continue the study by loading each of the remaining SPE cartridges with their respective sample load volumes.

1.0 mL Load Volume

On the second cartridge from the left in Figure 138, 1.0 mL of sample was loaded. Note how both dyes migrate further down the sorbent bed with larger sample volume being loaded. Again, some of the yellow dye has separated away from the rest of the sample, which still appears green. It appears green because some yellow dye is still present at the inlet, from the last part of the sample volume to be loaded. However, none of the yellow dye has flowed out of the cartridge when the loading step stops at 1.0 mL. The liquid that exits the SPE cartridge is collected and labeled.

2.0 mL Load Volume [Intended]

The process continues with the third cartridge, which has 2.0 mL of sample loaded. Here, some of the yellow dye elutes from the cartridge during the loading step. This is due to the sample volume being large enough and the k for yellow dye being low enough to allow some of the yellow dye to travel all the way through the cartridge during the loading step. The cartridge has no more capacity for the yellow dye because of the chromatographic conditions and the volume of sample that was loaded. Note that the rest of the sample, which appears greenish/blue, has traveled even further down the sorbent bed as larger amounts of blue and yellow dyes are loaded. Also note that none of the blue dye has eluted from the cartridge due to its higher k value. The liquid that exits the SPE cartridge appears yellow, and is collected and labeled. However, remember that it does contain some of the yellow dye. This will affect the % recovery calculation later.

3.0 mL Load Volume

On the fourth SPE cartridge, 3.0 mL of sample is loaded. The sorbent bed has become almost fully colored with a bluish/green from the sample matrix. Yellow dye can be seen eluting from the cartridge, however, no blue dye is eluted from the cartridge while the loading step was being performed. This indicates that the cartridge still has enough capacity for the blue dye at 3.0 mL of sample matrix volume. The liquid that exits the SPE cartridge is collected and labeled. However, remember that it also contains even more of the yellow dye. This will affect the % recovery calculation later.

4.0 mL Load Volume

The fifth SPE cartridge has 4.0 mL of sample loaded. Note that the sorbent bed is now fully colored bluish/green. A bluish/green color is also seen eluting from the cartridge during the loading step. This indicates that at 4.0 mL of sample, both the blue and yellow dyes have exceeded the capacity of the cartridge. The liquid that exits the SPE cartridge during the loading step appears green, and is collected and labeled. However, remember that it contains even more of the yellow dye and also some of the blue dye. This will affect the % recovery calculations for both the blue and yellow dyes.

5.0 mL Load Volume

And finally, for the sixth SPE cartridge, 5.0 mL of sample is loaded. The sorbent bed appears the same as was seen in the fifth cartridge. More of the bluish/green color is seen eluting from the cartridge during the loading step. This again indicates that the capacity of the cartridge to retain both of the dyes was exceeded, and as the sample is being loaded, some amount is not captured of both compounds of interest, because too much volume of sample is loaded onto the cartridge. The liquid that exits the SPE cartridge during the loading step appears green, and is collected and labeled. However, remember that it contains even more of the yellow dye and more of the blue dye. This will affect the % recovery calculation later for both the blue and yellow dyes.

The loading part of the breakthrough study is now complete. Remember, that we are trying to determine if this SPE cartridge has the capacity to retain the yellow and blue dyes, with a load volume of 2.0 mL of this sample.

Completing the Protocol

To complete the SPE protocol, there are two elution steps. The solvent for the elute 1 step is designed to completely elute all of the yellow dye only. A different, stronger solvent is used in elute 2, which will elute all of the blue dye only. For each of the six SPE cartridges, the appropriate volume of elute 1 solvent and elute 2 solvent is used as specified in the proposed method. What exits each cartridge is labeled as "Collect Elute 1—Yellow" and "Collect Elute 2—Blue" with the cartridge identity information.

All twelve elute samples are analyzed, and the mass recovered of each of the two dyes is determined. Next, the % recovery is calculated for both dyes for each of the six SPE cartridges as shown in the "Calculating Recovery" Section on page 69. Divide the mass recovered from the cartridge by the mass loaded and multiply by 100 for each dye. The mass loaded will be calculated from the dye concentration and the volume of sample loaded onto the SPE cartridge.

$$\% \text{ Recovery} = \frac{\text{Mass Recovered}}{\text{Mass Loaded}} \times 100$$

Remember that when the capacity of the SPE cartridge was exceeded, some amount of each of the dyes was not captured by the sorbent bed. They exited the cartridge during the load step and were never captured. Therefore, the mass recovered from the cartridge will be less by that amount. This is the breakthrough for that compound.

Plotting the Results

A simple X-Y bar graph plot of the % recovery for each dye versus the sample load volume is prepared. See Figure 140, which is repeated here to make it easier to follow the discussion.

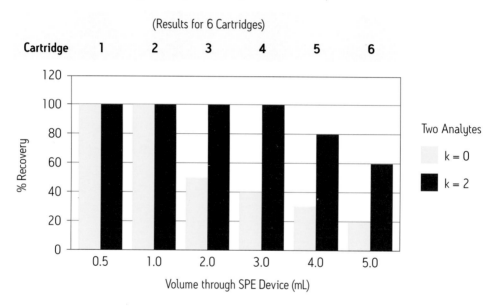

Figure 140: % Recovery vs. Load Volume

Each of the six sample volumes is plotted on the X-axis, and the % recovery values for the yellow and blue dyes are plotted using colored bars on the Y-axis. For the 0.5 and 1.0 mL sample volumes, 100% recovery is achieved for both dyes. Figure 141, which is a repeat of Figure 138, is provided here to make the discussion easier. Note that for both load volumes, neither of the dyes exits the cartridge during the load step. That indicates that all of the mass for both dyes was captured because the SPE cartridge had enough capacity. Therefore, during the elute 1—yellow step, all the yellow dye was recovered. The blue dye behaved the same way. Since all the mass that was loaded was then recovered in the elution step, the % recovery values were 100% for both dyes.

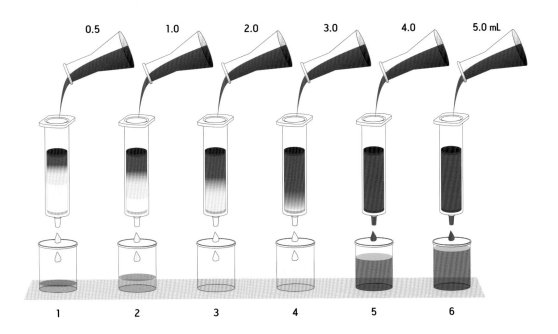

Figure 141: Loading Step

For the 2.0 mL sample volume, an interesting observation can be made. The % recovery value for the blue dye is still 100%, indicating that the SPE cartridge still has sufficient capacity for the blue dye. However, the yellow dye has 50% recovery, indicating that the SPE cartridge did not have sufficient capacity for the yellow dye under these chromatographic conditions with a 2.0 mL sample volume. In Figure 141, note that during the loading step, some of the yellow dye "broke through". It was seen in the liquid that exited the cartridge. From the % recovery calculation, and thinking about a mass balance for the yellow dye, 50% of the mass must have exited the SPE cartridge during the load step and was discarded.

For the 3.0 mL sample volume, the blue dye still has 100% recovery, while the yellow dye is down to ~40%. Figure 141 shows that the SPE cartridge still has just enough capacity for the blue dye and retains all of it. Note the recovery for the yellow dye is even less. Once the capacity of the cartridge is exceeded, as more volume of that compound is loaded, more simply passes through unretained, resulting in lower recovered mass values.

For the 4.0 mL sample volume, the recovery for the blue dye now drops below 100%, at ~80%, indicating that under these conditions, the cartridge size selected does not have sufficient capacity for 4.0 mL of sample volume. See Figure 139. As we can see in Figure 141, both the blue and yellow dyes exit the cartridge during the load step. Their recovered mass values will be less, resulting in poor % recovery.

The 5.0 mL sample volume experiment shows even lower % recovery values for both dyes: ~20% for yellow and ~60% for blue. This size and type of SPE cartridge does not have sufficient capacity for 5.0 mL of this sample.

Analyzing the Results of the Breakthough Study

Going back to the original question, "Can 2.0 mL of this green sample matrix be loaded onto the selected cartridge type and size for this SPE protocol?" The answer is "no," if we needed to recover both the yellow and blue dyes.

Based on the breakthrough study, the maximum volume that can be loaded is 1.0 mL, which will achieve 100% recovery for both the yellow and blue dye compounds. However, if the 2.0 mL capacity question was only regarding the blue dye, then the answer would be "yes." The experiment shows that up to 3.0 mL could be loaded with sufficient capacity for the blue dye.

In general, do not run a SPE cartridge near its capacity limit, because slight variations in the sample matrix or the actual precision in how the protocol is physically carried out will cause variations in % recoveries. For this sample, 2.0 mL would be a conservative value for the blue dye and 1.0 mL for the yellow and blue dyes.

If however, it was required to process a 2 mL sample volume for both dyes, and the SPE cartridge type and size could not be changed, then the sample solvent would have to change and be chromatographically weaker, thus increasing the k value for the yellow dye, which would allow more retention, thus more capacity.

If the sample volume, sample solvent, and SPE sorbent type could not be changed, then the SPE cartridge size could be increased to give more sorbent mass, which would provide more capacity. Changing from one size SPE cartridge sorbent mass to another size cartridge is called "scaling," which is discussed in the next section.

A breakthrough study is a very powerful part of the method development process to ensure proper and robust performance in a SPE protocol. Its importance should be clear.

Scaling of the Method Volumes and Sorbent Mass

To properly determine the correct SPE cartridge size [sorbent mass] for an application, breakthrough studies and mass balance determinations will be important. Moving to either a larger or smaller SPE cartridge size, in order to adapt an existing method for a desired change in the sample volume, will require that a scaling be performed on the sample volume as well as on the various solvent volumes used in the protocol.

For example, in the breakthrough study performed on page 160, it was found that the SPE cartridge size chosen did not have the capacity for both the blue and yellow compounds at the intended application sample volume of 2.0 mL. If it was required to process a 2.0 mL volume of the green sample, and the chromatographic conditions could not be changed, then a scaling calculation must be performed. The main scaling factor is determined by the sorbent mass ratio of the two SPE cartridges. It is calculated as follows:

$$\text{SPE Scaling Factor} = \frac{\text{New Sorbent Mass}}{\text{Original Sorbent Mass}}$$

This scaling factor is then used to calculate the new sample volume and each of the solvent volumes [i.e. wash and elute steps] to be used in the new protocol. Since we need twice the sample volume, we will need twice the sorbent.

[Scaling of the Method Volumes and Sorbent Mass]

By doubling the sorbent mass in a new cartridge, the scaling factor would be 2.0x. This value is then used to scale the sample volume and the wash and elution solvent volumes.

Table 10 shows specifics for the breakthrough study protocol that was performed on page 160.

Table 10: Scaling Information

	Original Conditions
Original sorbent mass	100 mg
Original sample volume	1.0 mL [based on breakthrough results]
Original elution volume [yellow]	1.5 mL
Original elution volume [blue]	2.0 mL

The breakthrough study indicated that the original cartridge only had the capacity for 1.0 mL of sample volume. Therefore, the intended sample volume is twice as large, indicating the need for a scaling factor of 2.0x. The new, scaled SPE conditions are shown on Table 11.

Table 11: Scaling for New Conditions

	Calculations	Scaled Conditions
Sorbent mass	100 mg x 2.0	200 mg
Sample volume	1.0 mL x 2.0	2.0 mL [intended]
Original elution volume [yellow]	1.5 mL x 2.0	3.0 mL
Original elution volume [blue]	2.0 mL x 2.0	4.0 mL

A 200 mg SPE cartridge will now have the capacity necessary for this application, and the scaled solvent volumes will provide proper chromatographic behavior in the protocol.

Sometimes you will need to scale in the opposite direction, to a reduced sample volume and save time and cost. In the following case [Table 12], the original conditions were:

Table 12: Scaling Information

	Original Conditions
Original sorbent mass	500 mg
Original sample volume	2.0 mL
Original wash solvent volume	3.0 mL
Original elute solvent volume	2.0 mL

[Scaling of the Method Volumes and Sorbent Mass]

The desired, new sample volume is to be 0.4 mL, indicating that a scaling factor of 0.2x will be required. New sorbent mass is calculated as follows:

$$\text{Scaling Factor} = 0.2 \, x = \frac{\text{New Sorbent Mass}}{500 \text{ mg}}$$

Solving for new sorbent mass = 0.2 x 500 mg = 100 mg

The new, scaled SPE protocol conditions are shown on Table 13.

Table 13: Scaling for New Conditions

	Calculations	Scaled Conditions
Sorbent mass	500 mg x 0.2	100 mg
Sample volume	2 mL x 0.2	0.4 mL [intended]
Wash solvent volume	3 mL x 0.2	0.6 mL
Elute solvent volume	2 mL x 0.2	0.4 mL

Remember to Adjust for Time

By following this scaling process, the performance of the new SPE protocol will meet the desired goals. Now that you have the correct volumes of liquids, make sure to adjust for the time for each step in order to maintain the proper linear velocity, as discussed in the "Key Terms and Calculations" section on page 66. This is especially critical when the internal bed diameter of the cartridges changes.

Troubleshooting

[Troubleshooting]

Tables 14 and 15 will detail some of the most common problems that occur in SPE methods. The information is segmented into three parts:

1. Problems encountered in the method development process [Table 14]
2. Problems with existing methods being executed routinely in the laboratory [Table 15]
3. General problems in sample preparation

Many problems encountered when creating SPE methods result from not properly performing the mass balance and breakthrough studies, as described in the "Method Development" section. Problems are usually related to poor recovery values or poor robustness as the method is carried out over time. Additionally, the calculations for % recovery, matrix interference effects [for mass spectrometry applications], and hold-up volume must be performed as described in the "Key Terms and Calculations" section. Variations in how the calculations are performed can lead to differences in results, especially if different analysts are involved.

Table 14: Problems Encountered in the Method Development Process

Pass-Through Strategy	
Problem	**Possible Solution**

Poor recovery for unretained [k = 0] compound of interest	• Perform mass balance and breakthrough study—if not done previously
	• Determine sample volume [or load volume] and the hold-up volume
	  - If load volume is greater, combine the "collect load" with the "collect elute" steps
	  - If load volume is less, ensure that the pure sample elution solvent is strong enough to obtain k = 0 and that an adequate volume is used to fully elute compound of interest
	• Use proper technique in calculating % recovery

Interferences eluting with compound of interest	• Check for flow rate control during load step to make sure linear velocity is not too fast • Reduce the volume of elution sample solvent • Ensure cartridge is properly wetted
Capture Strategy	
Problem	**Possible Solution**
Poor recovery for retained compound [initial k = high]	• Perform mass balance and breakthough study if not done previously • Ensure that sample solvent is weak enough to obtain high k value for complete capture • Choose more retentive sorbent to increase k value • Ensure sample temperatures are not too high, since this will result in less retention • Ensure breakthrough study completed to match SPE cartridge capacity to sample volume • Ensure elution step is complete with strong enough solvent, creating k = 0 and sufficient elution solvent volume • Ensure proper conditioning and equilibration steps using reversed-phase sorbents [no drying out effect/de-wetting] • Check for flow control during load, wash, and elute steps to make sure linear velocity is not too fast thereby causing poor initial capture and release

Table 14 continued.

Poor recovery for a retained, basic compound on a silica-based sorbent/particle	• Ensure that solvent pH is lowered to shut off unanticipated cation-exchange retention mechanism with negatively charged, ionized silanol groups on sorbent/particle surface. See page 169. • Use Oasis HLB copolymer sorbent to avoid risk of unanticipated cation-exchange mechanism
Interferences eluting with compound of interest	• Perform a mass balance • Ensure wash solvent volume and strength are strong enough to remove all of the less retained interferences • Ensure elution solvent is not too strong and proper flow control is maintained • Reduce the volume of elution solvent • Consider different mode of chromatography for SPE relative to the LC conditions. For example, use an orthogonal approach with an ion-exchange SPE method and a reversed-phase LC method
Ion suppression or ion enhancement in reversed-phase LC/MS	• Use a combined ion-exchange and reversed-phase sorbent, such as Oasis MCX, in an SPE protocol to provide cleaner extract

Table 15: Problems Encountered in Executing Existing methods

Pass-Through Strategy	
Problem	**Possible Solution**
Variable recoveries for an unretained compound of interest [k = 0]	• Check that mass balance was performed in methods development—if not, one needs to be performed • Check that sample load volume was correct • Ensure elution solvent volume and flow rate control are correct • Check that sample matrix has not changed, i.e. source or lot variation • Check that sample solvent is correct • Check that compound is not labile [temperature, light, pH] • Check that vial quality [glass] has not changed which can cause adsorption or degradation problems
Interferences eluting with compound of interest	• Check for flow control during load step to make sure linear velocity is not too fast • Ensure proper conditioning and equilibration steps using reversed-phase sorbents [no drying out effect/de-wetting]
Capture Strategy	
Problem	**Possible Solution**
Poor or variable recoveries for retained compound [initial k = high]	• Ensure proper conditioning and equilibration steps using reversed-phase sorbents [no drying out effect/de-wetting] • Check for flow rate control during load, wash and elution steps to make sure linear velocity is not too fast • Check that mass balance was performed in methods development if not, one needs to be performed • Check that compound is not labile [temperature, light, pH] • Check that vials glass quality has not changed, which can cause adsorption or degradation problems • Check for quality problems with SPE cartridges [vendor assessment]

[Troubleshooting]

General Problem Sources in Sample Preparation

Analysts performing SPE methods need proper training to ensure that the protocol is being performed correctly at all times. Variations in the execution of the protocol can result in poor performance reproducibility. The following areas are frequently the source of problems in sample preparation.

Temperature Control

In some applications, temperature control is critical. Make sure that all liquids are at the specified temperature before performing the method. Please remember that SPE is a chromatographic system, therefore temperature must be controlled to obtain consistent performance. As the temperature goes up, retention [capture] goes down, resulting in less capacity.

Flow Rate Control

Flow rate control is critical for robust performance in many SPE methods. The protocol should specify the appropriate flow rates for load, wash, and elute steps. Make sure that positive pressure, or vacuum-based sources used to develop liquid flow are set properly. Methods performed using vacuum manifolds must maintain specified vacuum levels in order to maintain consistent and proper flow rates.

pH Control

pH control is required to ensure the proper retention [capture] for acids and bases. Since their ionized forms have much lower k values, you will achieve much lower capture [retention] when acids are in high pH solutions and bases are in low pH solutions. The retention map for a pure reversed-phase, Oasis copolymer sorbent, shown below, diagrams this behavior.

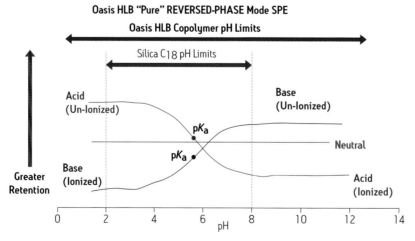

Note: Un-ionized form will have more retention. When ionized, the compound will be more polar and elute faster from the reversed-phase sorbent.

Figure 142: Retention Map for Pure Reversed-Phase Sorbent
[Also see Figure 26, and the discussion on page 43]

[Troubleshooting]

The term "pure" reversed-phase retention indicates that the sorbent does not have any other retention mechanism present, such as cation exchange. Some silica based reversed-phase sorbents can create a negatively charged surface [cation exchange] when their silanol groups deprotonate as the pH goes up. [Figure 143]

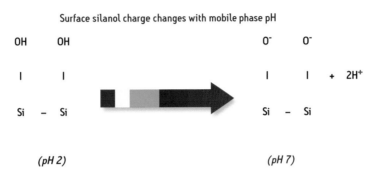

Figure 143: Effect of pH on Silica-Based SPE Sorbents

This will increase the retention for charged bases [+] because opposites attract [cation exchange], but will decrease the retention for charged acids [-] because likes repel. The retention map shown in Figure 144 is for a typical silica-based C_{18} SPE sorbent which can behave as a weak cation exchanger as pH approaches pH 5–7.

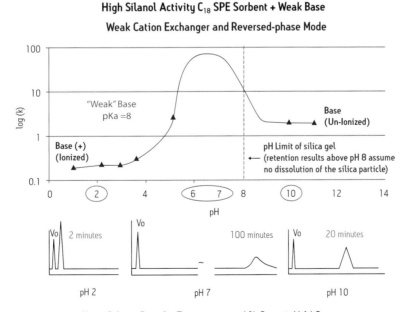

Figure 144: Retention Map for Silica C_{18} Sorbent

[Troubleshooting]

Compare the dramatic difference in retention behavior of the weak base in Figures 142 and 144. The cation exchange mechanism is activated when the base is positively charged and the sorbent is negatively charged, with a significant retention effect from pH 5 to pH 8. The map above pH 8 is theoretical, since the silica would begin dissolving in a high pH solution. As the "chromatograms" on the bottom of Figure 144 show, it is very hard to elute the weak base when the cation exchange is activated.

We can now modify the charge state of sorbents listing to include silica-based reversed-phase sorbents. [Figure 145]

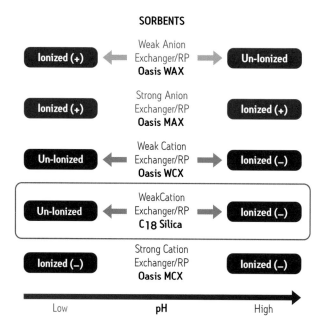

Figure 145: Sorbent Charge States

For applications where silica-based sorbents are used, this could cause poor recoveries for bases if the pH is not properly controlled especially during the load and wash and elution steps. [Figure 146]

[Troubleshooting]

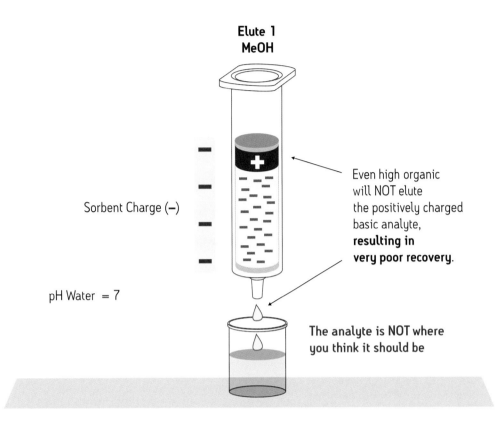

Figure 146: Poor Recovery of Bases Using Silica-Based Sorbents When pH is Not Controlled.

In this example, even when 100% MeOH is used during the elution step, the base does not elute. This is due to the significant retention form the cation-exchange mechanism developed because the sorbent is now negatively charged and the base is positively charged.

Here, during the elution step using 100% MeOH, the base does not elute due to the significant retention from cation exchange.

Pretreatment and Extraction

Sample pre-treatment and extraction studies are important for the long-term robustness of an SPE method. Processing of the original source of a solid sample, such as homogenization and centrifugation of foods, must be defined and followed prior to performing SPE.

In addition, an extraction study must be carried out to identify the conditions required to obtain the compounds of interest into the actual liquid sample that will be processed in the SPE protocol. For example, specific solvents or solvent mixtures, time, temperature, and centrifugation/sonication conditions must be developed to properly extract the compounds of interest from the original sample.

Certain drugs in biological fluids, such as plasma, can be bound to the proteins that are present in the sample matrix. When those interference proteins are washed away in the SPE protocol, the bound drug will also be washed away, resulting in poor recovery performance. Therefore, specific pretreatment of the sample matrix with acidic or basic solutions will be required to release the drug from the proteins before the SPE protocol is carried out.

[Troubleshooting]

Sample Matrix

Another common problem results from variations in the sample matrix itself. For example, human plasma from different patients can have different sets of interferences, which can cause performance problems for the SPE method. In addition, plasma samples from different species [for example moving from dog plasma to human] will sometimes show significant variations in interferences. SPE methods may have to be modified.

Vial Condition

The quality and cleanliness of the actual sample vial used in the analytical instrument can be a source of variability. Contamination is one major concern, especially today with the very high sensitivity requirements for modern methods. In addition, different analytical instrumentation, such as mass spectrometers, have specific needs for cleanliness. Your vial vendor has to be selected carefully since the vial is the last container for your sample before it is analyzed. If the vendor supplying the vials changes the process, or a new vial vendor is chosen, different levels and/or types of contamination can be introduced. Variations in contamination can come from the glass or polyethylene vial surface or the septa itself. In some cases, the contamination actually comes from the vial packaging, which can outgas contaminating compounds onto the vial surfaces.

Variations in the actual vial surface chemistry can also greatly impact the compounds of interest themselves. Several problems have been documented, which include the degradation of the compound of interest as well as the binding of the compound of interest to the vial wall. Both of these will cause analytical problems.

Sample Solvent

The sample solvent could also be responsible for degradation of the compound of interest. For samples that will be stored for extended periods of time, make sure that no significant degradation is occurring due to the solvent itself.

Appendix: Glossary of SPE & LC Terms

[Glossary]

Alumina

A porous, particulate form of aluminum oxide [Al_2O_3] used as a stationary phase in normal-phase adsorption chromatography. Alumina has a highly active basic surface; the pH of a 10% aqueous slurry is about 10. It is successively washed with strong acid to make neutral and acidic grades [slurry pH 7.5 and 4, resp.]. Alumina is more hygroscopic than silica. Its activity is measured according to the Brockmann scale for water content; e.g., Activity Grade I contains 1% H_2O.

H. Brockmann and H. Schodder, Ber. **74**: 73 [1941].

Analyte [see Constituent and Matrix]

The component or constituent of a sample to be analyzed. The analyte is found within the sample matrix.

Baseline*

The portion of the chromatogram recording the detector response when only the mobile phase emerges from the column.

Breakthrough Study

A study that is performed during method development to determine either the optimum size of SPE device, or the maximum amount of sample that can be loaded onto a chosen device, under a specified set of conditions. A range of sample volumes is loaded onto a set of SPE devices, and the % recovery values for each analyte are plotted to determine the maximum amount of sample that can be loaded onto the SPE device.

Cartridge

A type of column, without end fittings, that consists simply of an open tube wherein the packing material is retained by a frit at either end. SPE cartridges may be operated in parallel on a vacuum-manifold. HPLC cartridges are placed into a cartridge holder that has fluid connections built into each end. Cartridge columns are easy to change, less expensive, and more convenient than conventional columns with integral end fittings.

*Indicates a definition adapted from IUPAC sources, including the IUPAC Gold Book, available online at: http://goldbook.iupac.org/index.html and: Nomenclature for sampling in analytical chemistry, Pure Appl. Chem. 62: 1200 [1990], Nomenclature for automated and mechanised analysis, Pure Appl. Chem. 61: 1660 [1989], L.S. Ettre, Nomenclature for Chromatography, Pure Appl. Chem. 65: 819-872 [1993], ©1993 IUPAC; an updated version of this comprehensive report is available in the Orange Book, Chapter 9: Separations [1997] at: <http://www.iupac.org/publications/analytical_compendium>

[Glossary]

Chromatogram*

A graphical or other presentation of detector response or other quantity used as a measure of the concentration of the analyte in the effluent versus effluent volume or time. In planar chromatography [e.g., thin-layer chromatography or paper chromatography], chromatogram may refer to the paper or layer containing the separated zones.

Chromatography*

A dynamic physicochemical method of separation in which the components to be separated are distributed between two phases, one of which is stationary [the stationary phase] while the other [the mobile phase] moves relative to the stationary phase.

Column Volume* [Vc]

The geometric volume of the part of the tube that contains the packing [internal cross-sectional area of the tube multiplied by the packed bed length, L]. The interparticle volume of the column, also called the interstitial volume, is the volume occupied by the mobile phase between the particles in the packed bed [see Retention Time]. The void volume [Vo] is the total volume occupied by the mobile phase, i.e. the sum of the interstitial volume and the intraparticle volume [also called pore volume].

Continuous Gradient [see Gradient]

Constituent*

Chemical species present in a system; often called a component, although the term component has a more restricted meaning in physical chemistry.

Detector* [see Sensitivity]

A device that indicates a change in the composition of the eluent by measuring physical or chemical properties [e.g., UV/visible light absorbance, differential refractive index, fluorescence, or conductivity]. If the detector's response is linear with respect to sample concentration, then, by suitable calibration with standards, the amount of a component may be quantitated. Often, it may be beneficial to use two different types of detectors in series. In this way, more corroboratory or specific information may be obtained about the sample analytes. Some detectors [e.g., electrochemical, mass spectrometric] are destructive; i.e., they effect a chemical change in the sample components. If a detector of this type is paired with a non-destructive detector, it is usually placed second in the flow path.

Display

A device that records the electrical response of a detector on a computer screen in the form of a chromatogram. Advanced data recording systems also perform calculations using sophisticated algorithms, e.g., to integrate peak areas, subtract baselines, match spectra, quantitate components, and identify unknowns by comparison to standard libraries.

Distribution Constant [see Partition Ratio]

De-wetting or the Drying out Effect

A condition in reversed-phase chromatography that occurs when the pores of the sorbent [particle] are not wetted by a pure aqueous sample solvent or mobile phase. Since the pores are dry, analytes are not captured or retained as expected, resulting in poor recovery. This occurs if the conditioning step with organic solvent and equilibration step with sample solvent are not performed, or poorly controlled before the sample is loaded onto the SPE device.

Efficiency [H, see Plate Number, Resolution, Sensitivity, Speed]

A measure of a column's ability to resist the dispersion of a sample band as it passes through the packed bed. An efficient column minimizes band dispersion or bandspreading. Higher efficiency is important for effective separation, greater sensitivity, and/or identification of similar components in a complex sample mixture

Nobelists Martin and Synge, by analogy to distillation, introduced the concept of plate height [H, or H.E.T.P., height equivalent to a theoretical plate] as a measure of chromatographic efficiency and a means to compare column performance. Pre-dating HPLC and UPLC technology, they recognized that a homogeneous bed packed with the smallest possible particle size [requiring higher pressure] was key to maximum efficiency. The relation between column and separation system parameters that affect bandspreading was later described in an equation by van Deemter.

Chromatographers often refer to a quantity that they can calculate easily and directly from measurements made on a chromatogram, namely plate number [N], as efficiency. Plate height is then determined from the ratio of N to the length of the column bed [H = L/N] methods of calculating N are shown in Figure 147. It is important to note that calculation of N or H using these methods is correct only for isocratic conditions and cannot be used for gradient separations.

A.J.P. Martin and R.M. Synge, Biochem. J. 35: 1358–1368 [1941]

J.J. van Deemter, F. J. Zuiderweg and A. Klinkenberg Chem. Eng. Sci. 5: 271–289 [1956]

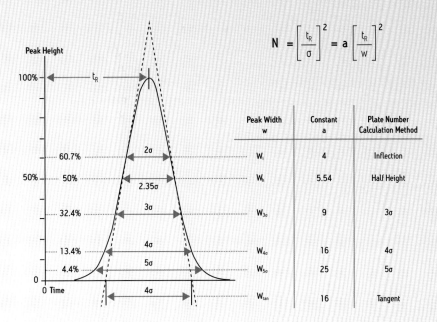

Figure 147: Methods for Calculating Plate Number [N]

Eluate

The portion of the eluent that emerges from the column or SPE cartridge outlet containing analytes in solution. In analytical LC, the eluate is examined by the detector for the concentration or mass of analytes. In preparative LC, the eluate is collected continuously in aliquots at uniform time or volume intervals, or discontinuously only when a detector indicates the presence of a peak of interest. These fractions are subsequently processed to obtain purified compounds. For SPE, the eluate is collected at specific steps in the method so that further analysis can be performed.

Eluent

The Mobile Phase [see Elution Chromatography].

Eluotropic Series

A list of solvents ordered by elution strength with reference to specified analytes on a standard sorbent. Such a series is useful when developing both isocratic and gradient elution methods. Trappe coined this term after showing that a sequence of solvents of increasing polarity could separate lipid fractions on alumina. Later, Snyder measured and tabulated solvent strength parameters for a large list of solvents on several normal-phase LC sorbents. Neher created a very useful nomogram by which equi-eluotropic [constant elution strength] mixtures of normal-phase solvents could be chosen to optimize the selectivity of TLC separations.

W. Trappe, Biochem. Z. 305: 150 [1940]
L. R. Snyder, Principles of Adsorption Chromatography, Marcel Dekker [1968], 192–197
R. Neher in G.B. Marini-Bettòlo, ed., Thin-Layer Chromatography, Elsevier [1964] pp. 75–86.

A typical normal-phase eluotropic series would start at the weak end with non-polar aliphatic hydrocarbons, e.g., pentane or hexane, then progress successively to benzene [an aromatic hydrocarbon], dichloromethane [a chlorinated hydrocarbon], diethyl ether, ethyl acetate [an ester], acetone [a ketone], and, finally, methanol [an alcohol] at the strong end.

Elute* [verb]

To chromatograph by elution chromatography. The process of elution may be stopped while all the sample components are still on the chromatographic bed [planar thin-layer or paper chromatography] or continued until the components have left the chromatographic bed [column chromatography].

Note: The term elute is preferred to develop [a term used in planar chromatography], to avoid confusion with the practice of method development, whereby a separation system [the combination of mobile and stationary phases] is optimized for a particular separation.

Elution Chromatography*

A procedure for chromatographic separation in which the mobile phase is continuously passed through the chromatographic bed. In HPLC, once the detector baseline has stabilized and the separation system has reached equilibrium, a finite slug of sample is introduced into the flowing mobile phase stream. Elution continues until all analytes of interest have passed through the detector.

Elution Strength

A measure of the affinity of a solvent relative to that of the analyte for the stationary phase. A weak solvent cannot displace the analyte, and therefore the analyte is strongly retained on the stationary phase. A strong solvent may totally displace all the analyte molecules and carry them through the column unretained. To achieve a proper balance of effective separation and reasonable elution volume, solvents are often blended to set up an appropriate competition between the phases, thereby optimizing both selectivity and separation time for a given set of analytes [see Selectivity].

Dipole moment, dielectric constant, hydrogen bonding, molecular size and shape, and surface tension may give some indication of elution strength. Elution strength is also determined by the separation mode. An eluotropic series of solvents may be ordered by increasing strength in one direction under adsorption or normal-phase conditions; that order may be nearly opposite under reversed-phase partition conditions. See Figure 148.

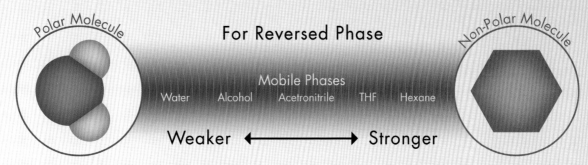

Figure 148: Elution Strength for Reversed Phase

Fluorescence Detector

Fluorescence detectors excite a sample with a specified wavelength of light. This causes certain compounds to fluoresce and emit light at a higher wavelength. A sensor, set to a specific emission wavelength and masked so as not to be blinded by the excitation source, collects only the emitted light. Often analytes that do not natively fluoresce may be derivatized to take advantage of the high sensitivity and selectivity of this form of detection, e.g., AccQ•Tag™ derivatization of amino acids.

Flow Rate*

The volume of mobile phase passing through the column or SPE cartridge in unit time. In LC systems, the flow rate is set by the controller for the solvent delivery system [pump]. Flow rate accuracy can be checked by timed collection and measurement of the effluent at the column outlet. Since a solvent's density varies with temperature, any calibration or flow rate measurement must take this variable into account. Most accurate determinations are made, when possible, by weight, not volume.

Uniformity [precision] and reproducibility of flow rate is important to many LC techniques, especially in separations where retention times are key to analyte identification, or in gel-permeation chromatography where calibration and correlation of retention times are critical to accurate molecular-weight-distribution measurements of polymers.

Often, separation conditions are compared by means of linear velocity, not flow rate. The linear velocity is calculated by dividing the flow rate by the cross-sectional area of the column. While flow rate is expressed in volume/time [e.g., mL/min], linear velocity is measured in length/time [e.g., mm/sec]. or SPE cartridge

In SPE, controlling the vacuum or positive displacement pumping system is critical to achieve reproducible flow rates through the SPE devices. The loading, washing, and elution steps require controlled flow rates.

Gel-Permeation Chromatography*

Separation based mainly upon exclusion effects due to differences in molecular size and/or shape. Gel-permeation chromatography [GFC] and gel-filtration chromatography [GFC] describe the process when the stationary phase is a swollen gel. Both are forms of Size-exclusion chromatography [SEC]. Porath and Flodin first described gel-filtration using dextran gels and aqueous mobile phases for the size-based separation of biomolecules. Moore applied similar principles to the separation of organic polymers by size in solution using organic-solvent mobile phases on porous polystyrene-DVB polymer gels.

J. Porath, P. Flodin, Nature 183: 1657–1659 [1959] J.C. Moore, U.S. Patent 3,326,875 [filed Jan. 1963; issued June 1967]

Gradient

The change over time in the relative concentrations of two [or more] miscible solvent components that form a mobile phase of increasing elution strength. A step gradient is typically used in solid-phase extraction; here the eluent composition is changed abruptly from a weaker mobile phase to a stronger mobile phase. It is even possible, by drying the SPE sorbent bed in between steps, to change from one solvent to another immiscible solvent.

A continuous gradient is typically generated by a low- or high-pressure mixing system [see Figures 149 and 150] according to a pre-determined curve [linear or non-linear] representing the concentration of the stronger solvent B in the initial solvent A over a fixed time period. A hold at a fixed [isocratic] solvent composition can be programmed at any time point within a continuous gradient. At the end of a separation, the gradient program can also be set to return to the initial mobile phase composition to re-equilibrate the column in preparation for the injection of the next sample. Sophisticated LC systems can blend as many as four or more solvents [or solvent mixtures] into a continuous gradient.

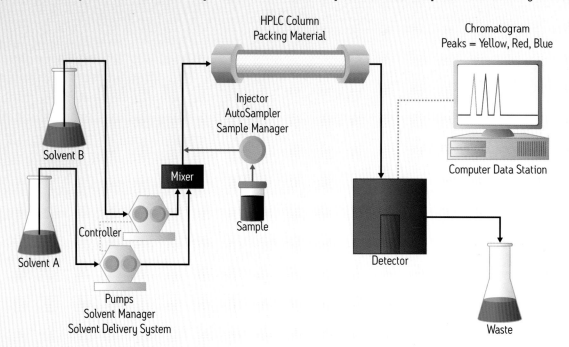

Figure 149: High Pressure Gradient System

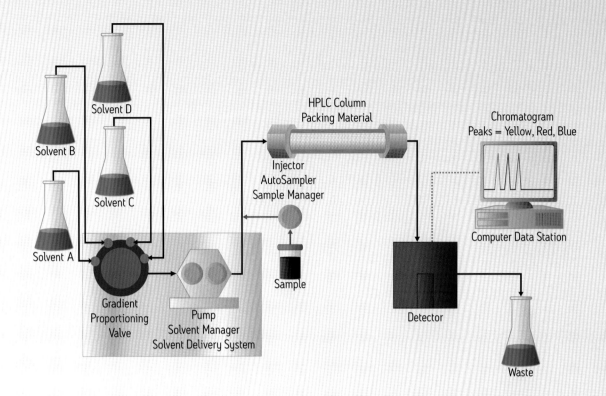

Figure 150: Low Pressure Gradient System

Hold-Up Volume [time], VM [tM] [in column chromatography]

The volume of the mobile phase [or the corresponding time] required to elute a component, the concentration of which in the stationary phase is negligible compared to that in the mobile phase. In other words, this component is not retained at all by the stationary phase. Thus, the hold-up volume [time] is equal to the retention volume [time] of an unretained compound. The hold-up volume [time] includes any volumes contributed by the sample injector, the detector and connectors.

For SPE devices, the hold-up volume includes the interstitial volume between sorbent particles, the pore volume of the particles, the liquid included in the inlet and outlet filters and the liquid in the outlet connector. See Figure 151.

[Glossary]

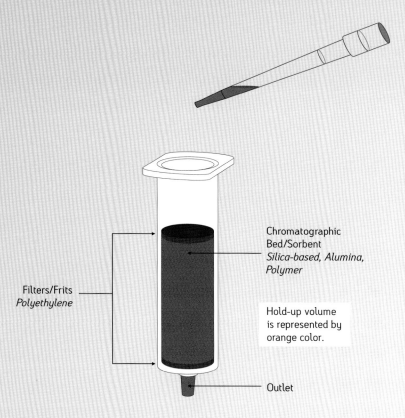

Figure 151: Hold-Up Volume—SPE Cartridge

Injector [Autosampler, Sample Manager]

A mechanism for introducing [injecting] accurately and precisely a discrete, predetermined volume of a sample solution into the flowing mobile phase stream. The injector can be a simple manual device, or a sophisticated autosampler that can be programmed for unattended injections of many samples from an array of individual vials or wells in a predetermined sequence. Sample compartments in these systems may even be temperature controlled to maintain sample integrity over many hours of operation.

Most modern injectors incorporate some form of syringe-filled sample loop that can be switched on- or offline by means of a multi-port valve. A well-designed, minimal-internal-volume injection system is situated as close to the column inlet as possible and minimizes the spreading of the sample band. Between sample injections, it is also capable of being flushed to waste by mobile phase, or a wash solvent, to prevent carryover [contamination of a sample by a previous one].

Samples are best prepared for injection, if possible, by dissolving them in the mobile phase into which they will be injected; this may prevent issues with separation and/or detection. If another solvent must be used, it is desirable that its elution strength be equal to or less than that of the mobile phase. It is often wise to mix a bit of a sample solution with the mobile phase offline to test for precipitation or miscibility issues that might compromise a successful separation.

[Glossary]

Inlet

The end of the column or SPE device bed where the mobile phase stream and sample enter. A porous, inert frit retains the packing material and protects the sorbent bed inlet from particulate contamination. Good HPLC practice dictates that samples and mobile phases should be particulate-free; this becomes imperative for small-particle UPLC columns whose inlets are much more easily plugged. If the column bed inlet becomes clogged and exhibits higher-than-normal backpressure, sometimes reversing the flow direction while directing the effluent to waste may dislodge and flush out sample debris that sits atop the frit. If the debris has penetrated the frit and is lodged in the inlet end of the bed itself, then the column has most likely reached the end of its useful life.

Ion-Exchange Chromatography* [see section: Separations Based on Charge]

This separation mode is based mainly on differences in the ion-exchange affinities of the sample components.

Separation of primarily inorganic ionic species in water or buffered aqueous mobile phases on small-particle, superficially porous, high-efficiency, ion-exchange columns followed by conductometric or electrochemical detection is referred to as ion chromatography [IC].

Isocratic Elution*

A procedure in which the composition of the mobile phase remains constant, during the elution process.

k [Retention Factor]

The value developed by the competition of the sorbent and the solvent for the analyte. If $k = 0$ then the analyte is not retained by the sorbent. If $k =$ much greater than 0 then the analyte is strongly retained by the sorbent. Also see "Retention Factor".

Laboratory Sample*

The sample or subsample[s] sent to or received by the laboratory. When the laboratory sample is further prepared [reduced] by subdividing, mixing, grinding or by combinations of these operations, the result is the test sample. When no preparation of the laboratory sample is required, the laboratory sample is the test sample. A test portion is removed from the test sample for the performance of the test or for analysis. The laboratory sample is the final sample from the point of view of sample collection but it is the initial sample from the point of view of the laboratory. Several laboratory samples may be prepared and sent to different laboratories or to the same laboratory for different purposes. When sent to the same laboratory, the set is generally considered as a single laboratory sample and is documented as a single sample.

Linear Velocity [see Flow Rate]

Liquid Chromatography* [LC]

A separation technique in which the mobile phase is a liquid. Liquid chromatography can be carried out either in a column or on a plane [TLC or paper chromatography]. Modern liquid chromatography utilizing smaller particles and higher inlet pressure was termed high-performance [or high-pressure] liquid chromatography [HPLC] in 1970. In 2004, ultra-pressure liquid chromatography dramatically raised the performance of LC to a new plateau [see UPLC Technology].

Mass Balance

In SPE, this is an experiment designed to determine where an individual analyte is located at each step in the protocol. It is used in combination with % recovery calculations to see if the analyte is fully or partially located in the SPE device, or has been fully or partially eluted during a load, wash or elution step.

Matrix

The components of the sample other than the analyte[s] or constituent[s] of interest.

Matrix Effect Calculation

This is an experiment to determine if any of the sample matrix components are causing response problems for the analyte of interest, in the mass spectrometer. These matrix components can either have no effect, or they can suppress or enhance the response of the analyte of interest.

Mobile Phase* [see Eluate, Eluent]

A fluid that percolates, in a definite direction, through the length of the stationary-phase sorbent bed. The mobile phase may be a liquid [liquid chromatography] or a gas [gas chromatography] or a supercritical fluid [supercritical-fluid chromatography (SPC)]. In gas chromatography the expression carrier gas may be used for the mobile phase. In elution chromatography, the mobile phase may also be called the eluent, while the word eluate is defined as the portion of the mobile phase that has passed through the sorbent bed and contains the compounds of interest in solution.

Normal-Phase Chromatography*

An elution procedure in which the stationary phase is more polar than the mobile phase. This term is used in liquid chromatography to emphasize the contrast to reversed-phase chromatography.

Partition Ratio, KD

The ratio of the concentration of a substance in a single definite form, A, in the extract to its concentration in the same form in the other phase at equilibrium, e.g. for an aqueous/organic system:

Notes:
1. Sometimes called the distribution constant; which is a good synonym. The terms distribution coefficient, distribution ratio, partition constant and extraction constant should not be used as synonyms for partition ratio.
2. The use of the inverse ratio [aqueous/organic] may be appropriate in certain cases, e.g. where the organic phase forms the feed, but its use in such cases should be clearly specified. The ratio of the concentration in the denser phase to the less dense phase is not recommended, as it can be ambiguous.

Peak* [see Plate Number]

The portion of a differential chromatogram recording the detector response while a single component is eluted from the column. If separation is incomplete, two or more components may be eluted as one unresolved peak. Peaks eluted under optimal conditions from a well-packed, efficient column, operated in a system that minimizes bandspreading, approach the shape of a Gaussian distribution. Quantitation is usually done by measuring the peak area [enclosed by the baseline and the peak curve]. Less often, peak height [the distance measured from the peak apex to the baseline] may be used for quantitation. This procedure requires that both the peak width and the peak shape remain constant.

Plate Number* [N, see Efficiency]

A number indicative of column performance [mechanical separation power or efficiency, also called plate count, number of theoretical plates, or theoretical plate number]. Plate number relates the magnitude of a peak's retention to its width [variance or bandspread]. In order to calculate a plate count, it is assumed that a peak can be represented by a Gaussian distribution [a statistical bell curve]. At the inflection points [60.7% of peak height], the width of a Gaussian curve is twice the standard deviation [s] about its mean [located at the peak apex]. As shown in Figure 152, a Gaussian curve's peak width measured at other fractions of peak height can be expressed in precisely defined multiples of s. Peak retention [retention volume, VR, or retention time, t_R] and peak width must be expressed in the same units, because N is a dimensionless number. Note that the 5s method of calculating N is a more stringent measure of column homogeneity and performance, as it is more severely affected by peak asymmetry. Computer data stations can automatically delineate each resolved peak and calculate its corresponding plate number.

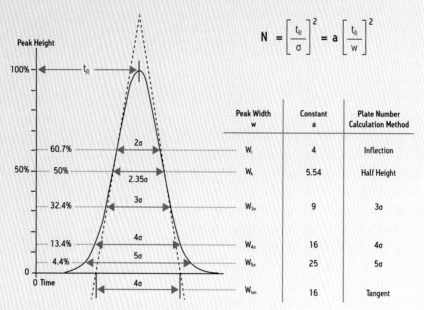

Figure 152: Methods for Calculating Plate Number [N]

Preparative Chromatography

The process of using liquid chromatography to isolate a compound in a quantity and at a purity level sufficient for further experiments or processes. For pharmaceutical or biotechnological purification processes, columns several feet in diameter can be used for multiple kilograms of material. For isolating just a few micrograms of a valuable natural product, an analytical HPLC column is sufficient. Both are preparative chromatographic approaches, differing only in scale.

[Glossary]

Recovery [%]

This is a determination of how much analyte was eluted from the SPE device, relative to how much analyte was initially loaded onto the device.

Resolution* [Rs, see Selectivity]

The separation of two peaks, expressed as the difference in their corresponding retention times, divided by their average peak width at the baseline. Rs = 1.2 indicates that two peaks of equal width are just separated at the baseline. When Rs = 0.6, the only visual indication of the presence of two peaks on a chromatogram is a small notch near the peak apex. Higher efficiency columns produce narrower peaks and improve resolution for difficult separations; however, resolution increases by only the square root of N. The most powerful method of increasing resolution is to increase selectivity by altering the mobile/stationary phase combination used for the chromatographic separation [see section on Chemical Separation Power].

$$RS = \frac{\sqrt{N}}{4} \left(\frac{\alpha-1}{\alpha}\right)\left(\frac{1+k_2}{k_2}\right)$$

N = Plate Count
α = Selectivity
k_2 = Retention factor for the 2nd peak

Reversed-Phase Chromatography*

An elution procedure used in liquid chromatography in which the mobile phase is significantly more polar than the stationary phase, e.g. a microporous silica-based material with alkyl chains chemically bonded to its accessible surface. Note: Avoid the incorrect term reverse phase.

Retention Factor* [k]

A measure of the time the sample component resides in the stationary phase relative to the time it resides in the mobile phase; it expresses how much longer a sample component is retarded by the stationary phase than it would take to travel through the column with the velocity of the mobile phase. Mathematically, it is the ratio of the adjusted retention time [volume] and the hold-up time [volume]: $k = t_R'/t_M$ [see Retention Time and Selectivity].

Note: In the past, this term has also been expressed as Partition Ratio, Capacity Ratio, Capacity Factor, or Mass Distribution Ratio and symbolized by k'.

Retention Time* [tR, see Column Volume]

The time between the start of elution [typically, in HPLC, the moment of injection or sample introduction] and the emergence of the peak maximum. The adjusted retention time, t_R', is calculated by subtracting from t_R the hold-up time [t_M, the time from injection to the elution of the peak maximum of a totally unretained analyte].

[Glossary]

Sample* [in analytical chemistry]

A portion of material selected from a larger quantity of material. The term needs to be qualified, e.g. bulk sample, representative sample, primary sample, bulked sample, test sample. The term 'sample' implies the existence of a sampling error, i.e. the results obtained on the portions taken are only estimates of the concentration of a constituent or the quantity of a property present in the parent material. If there is no or negligible sampling error, the portion removed is a test portion, aliquot or specimen. The term 'specimen' is used to denote a portion taken under conditions such that the sampling variability cannot be assessed [usually because the population is changing], and is assumed, for convenience, to be zero. The manner of selection of the sample should be prescribed in a sampling plan.

Sample* [in chromatography]

The mixture consisting of a number of components, the separation of which is attempted on the chromatographic bed as they are carried or eluted by the mobile phase.

Sample Components*

The chemically pure constituents of the sample. They may be unretained [i.e., not delayed] by the stationary phase, partially retained [i.e., eluted at different times], or retained permanently. The terms "elute" and "analyte" are also acceptable for a sample component.

Selectivity [Separation Factor, α]

A term used to describe the magnitude of the difference between the relative thermodynamic affinities of a pair of analytes for the specified mobile and stationary phases that comprise the separation system. The proper term is Separation Factor [α]. It equals the ratio of retention factors, k_2/k_1 [see Retention Factor]; by definition, α is always ≥ 1. If α = 1, then both peaks co-elute, and no separation is obtained. It is important in preparative chromatography, including SPE, to maximize α for highest sample loadability and throughput.

Sensitivity* [S]

The signal output per unit concentration or unit mass of a substance in the mobile phase entering the detector, e.g., the slope of a linear calibration curve [see Detector]. For concentration-sensitive detectors [e.g., UV/VIS absorbance], sensitivity is the ratio of peak height to analyte concentration in the peak. For mass-flow-sensitive detectors, it is the ratio of peak height to unit mass. If sensitivity is to be a unique performance characteristic, it must depend only on the chemical measurement process, not upon scale factors.

The ability to detect [qualify] or measure [quantify] an analyte is governed by many instrumental and chemical factors. Well-resolved peaks [maximum selectivity] eluting from high efficiency columns [narrow peak width with good symmetry for maximum peak height] as well as good detector sensitivity and specificity are ideal. Both the separation system interference and electronic component noise should also be minimized to achieve maximum sensitivity.

Solid-Phase Extraction [SPE]

A sample preparation technique that uses LC principles to isolate, enrich, and/or purify analytes from a complex matrix applied to a miniature chromatographic bed. Off-line SPE is done [manually or via automation] with larger particles in individual plastic cartridges or in micro-elution plate wells, using low positive pressure or vacuum to assist flow. On-line SPE is done with smaller particles in miniature HPLC columns using higher pressures and a valve to switch the SPE column on-line with the primary HPLC column, or off-line to waste, as appropriate.

SPE methods use step gradients [see Gradient] to accomplish bed conditioning, sample loading, washing, and elution steps. Samples are loaded typically under conditions where the k of important analytes is as high as possible, so that they are fully retained during loading and washing steps. Elution is then done by switching to a much stronger solvent mixture [see Elution Strength]. The goal is to remove matrix interferences and to isolate the analyte in a solution, and at a concentration, suitable for subsequent analysis.

Speed [see Efficiency, Flow Rate, Resolution]

A benefit of operating LC separations at higher linear velocities using smaller-volume, smaller-particle analytical columns, or larger-volume, larger-particle preparative columns. Order-of-magnitude advances in LC speed came in 1972 [with the use of 10-μm particles and pumps capable of delivering accurate mobile phase flow at 6000 psi], in 1976 [with 75-μm preparative columns operated at a flow rate of 500 mL/min], and in 2004 [with the introduction of UPLC technology—1.7-μm-particle columns operated at 15,000 psi].

High-speed analytical LC systems must not only accommodate higher pressures throughout the fluidics; injector cycle time must be short; gradient mixers must be capable of rapid turnaround between samples; detector sensors must rapidly respond to tiny changes in eluate composition; and data systems must collect the dozens of points each second required to plot and to quantitate narrow peaks accurately.

Together, higher resolution, higher speed, and higher efficiency typically deliver higher throughput. More samples can be analyzed in a workday. Larger quantities of compound can be purified per run or per process period.

See Beginner's Guide to UPLC book on list of Additional Reference Materials on page 208 for further reading.

Stationary Phase*

One of the two phases forming a chromatographic system. It may be a solid, a gel or a liquid. If a liquid, it may be distributed on a solid. This solid may or may not contribute to the separation process. The liquid may also be chemically bonded to the solid [bonded phase] or immobilized onto it [immobilized phase].

The expression chromatographic bed or sorbent may be used as a general term to denote any of the different forms in which the stationary phase is used.

The use of the term liquid phase to denote the mobile phase in LC is discouraged, in order to avoid confusion with gas chromatography where the stationary phase is called a liquid phase [most often a liquid coated on a solid support].

[Glossary]

Open-column liquid-liquid partition chromatography [LLC] did not translate well to HPLC. It was supplanted by the use of bonded-phase packings. LLC proved incompatible with modern detectors because of problems with bleed of the stationary-phase-liquid coating off its solid support, thereby contaminating the immiscible liquid mobile phase.

For SPE devices, the two most common stationary phases are built on silica particles, usually chemically bonded with a chromatographic ligand [C_{18}], or polymeric particles that can also be funtionalized for ion exchange behavior.

Step Gradient [see Gradient]

Unit [item/portion/individual]* [in analytical chemistry]

Each of the discrete, identifiable portions of material suitable for removal from a population as a sample or as a portion of a sample, and which can be individually considered, examined or tested, or combined. In the case of sampling bulk materials [or large packages], the units are increments, created by a sampling device. In the case of packaged materials, the unit may vary with the level of commercial distribution. For example, an individual piece of candy is the sampling unit at the consumer level; a package of individual pieces is the sampling unit at the retail level; a carton of packages is the sampling unit at the wholesale level; a pallet of cartons is the shipping unit at the distribution center level; and a truckload of pallets is the consignment unit at the manufacturing level. Before packaging, the bin containing the individual pieces would be the bulk lot [or batch] for sampling.

UPLC [Ultra Performance Liquid Chromatography] Technology

The use of a high-efficiency LC system holistically designed to accommodate sub-2 μm particles and very high operating pressure.

For more information, visit: http://www.waters.com/uplc

Appendix: Oasis Sorbent Technology for SPE

[Oasis Sorbent Technology for SPE]

Introduction

When SPE was first invented in the late 1970's, silica and alumina were primarily used as the substrates for the sorbents (packing materials), and they performed reasonably well. Over the decades, the chromatographic modes of reversed phase (e.g. C_{18}) and ion exchange became the most popular. However, it became clear over time that these sorbents suffered from several limitations including:

a) Poor water wettability resulting in % recovery problems requiring re-testing

b) Limited pH operating range (low pH only for silica-based materials)

c) Limited chromatographic capacity especially for polar analyte retention in reversed phase

d) Potential for unexpected silanol activity (cation exchange) resulting in poor recovery for charged bases and reduced capacity for charged acids

e) Poor batch-to-batch reproducibility.

Research was undertaken to develop a much more efficient performing family of sorbents designed specifically for SPE which addressed each one of these limitations. This resulted in a new, specially formulated patented co-polymeric particle, introduced in 1996, as Oasis Sorbent Technology. It is designed to provide the analytical scientist with a more dependable and robust level of performance, while making method development much simpler.

Even the name "Oasis" (a fertile, wet place in the desert) indicates that it was designed so that it doesn't suffer from the de-wetting (drying out) problems of traditional reversed-phase sorbents, thus providing more consistent sample results cartridge to cartridge. Some of the other advantages of this patented co-polymer particle are:

a) Greatly increased surface area (800 m^2/g vs 300 m^2/g) for more capacity so that less sorbent is needed resulting in higher analyte concentrations

b) Extended operating pH range from 1–14

c) Hydrophilic monomer provides enhance polar analyte capacity

d) No silica silanol groups present, resulting in fewer unexpected problems from ion exchange.

For more information on Oasis technology and applications, visit www.waters.com/oasis

[Oasis Sorbent Technology for SPE]

A Wide Selection of Oasis Chemistries

Oasis sorbents are available in 5 different SPE chemistries, providing a range of options for method development. The Oasis HLB sorbent is a macroporous copolymer made from a balanced ratio of two monomers, the lipophilic divinylbenzene and the hydrophilic N-vinylpyrrolidone. It provides reversed-phase capability with a special "polar hook" for enhanced capture of polar analytes and excellent wettability.

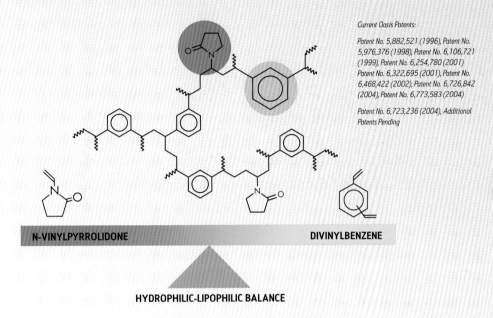

Current Oasis Patents:

Patent No. 5,882,521 (1996), Patent No. 5,976,376 (1998), Patent No. 6,106,721 (1999), Patent No. 6,254,780 (2001) Patent No. 6,322,695 (2001), Patent No. 6,468,422 (2002), Patent No. 6,726,842 (2004), Patent No. 6,773,583 (2004)

Patent No. 6,723,236 (2004), Additional Patents Pending

Specific Surface Area: 810 m²/g, Average Pore Diameter: 80Å
Total Pore Volume: 1.3 cm³/g, Average Particle Diameter: 30 mm or 60 mm

Figure 153: Unique Water-Wettable Oasis HLB Copolymer

[Oasis Sorbent Technology for SPE]

The other 4 Oasis chemistries—MCX, MAX, WCX, and WAX—are all derived from the Oasis HLB copolymer and feature a mixed-mode retention mechanism [both ion exchange and reversed phase], which can be manipulated very predictably for maximum selectivity and sensitivity. [Figure 154]

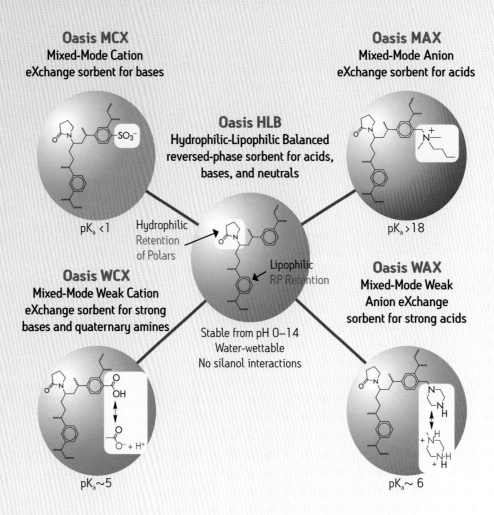

Figure 154: Oasis Sorbents

High and Consistent Recoveries

Oasis sorbents are water wettable, maintaining high retention and capacity for a wide spectrum of analytes. When a traditional SPE column runs dry, the sorbent pores dry out, the chromatographic retention [capture] of the analytes is reduced, resulting in poor recovery.

Silica-based C_{18} sorbents can easily dry out, especially on a vacuum manifold if a particular cartridge flows quickly and allows air to be drawn in. Oasis sorbents maintain proper wetting for more consistent performance [especially important for 96-well plates]. Even if air passes through, the Oasis pores do not dry out. [Figure 155]

Effect of Drying on Recovery—Oasis HLB Versus C_{18} Sorbents

No Impact of Sorbent Drying on HLB-High, Consistent Recovery

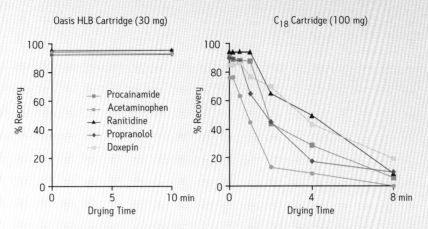

Oasis HLB 1 cc/30 mg and C_{18} 1 cc/100 mg Cartridges were Conditioned on a Waters Vacuum Manifold

Figure 155: Oasis HLB: No Impact of Sorbent Drying Out

When the methanol reached the top of the upper frit in each cartridge, vacuum was maintained for different times to vary the cartridge drying time. The SPE protocol was then continued. The data shown are the average of three replicate extractions.

The variable recoveries seen with the C_{18} sorbents, due to the drying out effect, are often the cause for "retests", reducing laboratory productivity. In some laboratories, 10% of samples are retests—this can be reduced using Oasis sorbents.

[Oasis Sorbent Technology for SPE]

Also, Oasis sorbents retain polar compounds far better than bonded silica SPE sorbents. Note the poor recovery of the polar analyte Acetaminophen on C_{18}. Oasis sorbents work especially well when you need to capture metabolites See Figure 156.

High Capacity Using Less Sorbent

When transferring methods from a C_{18}-bonded phase to Oasis products, keep in mind the greater capacity of the Oasis sorbent. The Oasis sorbent has 2–3x more surface area and shows a dramatic increase in k values compared to silica-based C_{18}. This reduces breakthrough potential. In addition, you may be able to use $^2/_3$ less sorbent than you would with C_{18} [30 mg of Oasis HLB gives equivalent capacity to 100 mg of C_{18}].

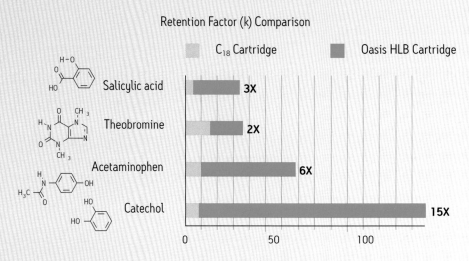

Figure 156: Higher Retention Means Greater Capacity, No Breakthrough

Exceptional Batch-to-Batch Reproducibility

Because of poor stability at pH extremes and relatively low ionic capacity, traditional silica-based, mixed-mode sorbents don't have long term, batch-to-batch reproducibility and, therefore, require reservations of specific lots of sorbent for large projects. Oasis sorbents have demonstrated excellent long-term, batch-to-batch reproducibility for over 15 years. As a result of careful process design and stringent quality controls, a new standard has been set in batch-to-batch and lot-to-lot reproducibility for SPE sorbents.

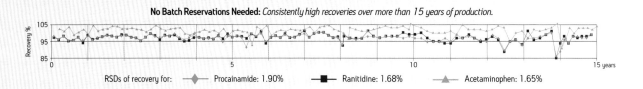

Figure 157: Batch-to-Batch Reproducibility of Oasis HLB Sorbent

Sorbent Amount and Solvent Selection for the Generic SPE Method

The suggested amount of sorbent in a cartridge or a plate required for your application is given in the Table 16 below. Remember, because of the increased capacity of the Oasis sorbents, you can use less sorbent than you would normally need if you used a silica-based packing. When converting from C_{18} silica-based sorbents to Oasis SPE sorbents, use approximately $^2/_3$ less Oasis sorbent [100 mg C_{18} sorbent = 30 mg Oasis sorbent].

Select the solvent used for the elution step based on the polarity of the analyte. The table to the right gives a selection of elution solvents and each solvent gives you different selectivity and elution strength.

Table 16: Capacity and Elution Volume of Oasis 96-Well Plates and Cartridges

Sorbent Per Device	Maximum Mass Capacity	Typical Sample Volumes	Elution Volume
2 mg [µElution Plate]*	60–400 µg	10–375 µL	25 µL**
5 mg*	0.15–1 mg	10–100 µL	≤ 150 µL
10 mg	0.35–2 mg	50–200 µL	≤ 250 µL
30 mg	1–5 mg	100 µL–1 mL	≥ 400 µL
60 mg	2–10 mg	200 µL–2 mL	≥ 800 µL

* Available only in 96-well plate formats
** µElution Plate requires no evaporation step

Table 17: Tips for Selecting Elution Solvents for the Generic SPE Method [1-D]† The Elution Solvent is Selected Based on Polarity of Analyte.

Solvent	Solvent Type	Relative Elution Strength††	Comments
Methanol	proton donor	1.0	disrupts H-bonding
Acetonitrile	dipole-dipole	3.1	medium polarity drugs
Tetrahydrofuran	dipole-dipole	3.7	medium polarity drugs
Acetone	dipole-dipole	8.8	medium polarity drugs
Ethyl Acetate	dipole-dipole	high	nonpolar drugs and GC compatible
Methylene Chloride	dipole-dipole	high	nonpolar drugs and GC compatible

† When using solvents other than methanol, add 10–30% [of proton donor solvent such as methanol] to disrupt H-bonding on the Oasis HLB sorbent.

†† High-Purity Solvent Guide. Burdick & Jackson Laboratories, Inc. Solvent Properties of Common Liquids, L.R. Snyder, J. Chromatogr., 92, 223 [1974]; J. Chromatogr. Sci. 16, 223 [1978]

Simplified Method Development Protocol

Oasis 2x4 Methodology for Cartridge and Standard 96-Well Plates

The Oasis 2x4 Method is a simple, logical approach to the selection of an SPE sorbent and protocol. Two protocols and four sorbents provide the flexibility to extract acids, bases, and neutrals with high recoveries while removing matrix components that may interfere with analysis.

Follow the simple steps outlined in this flow chart to achieve high recoveries and the cleanest extracts:

- Characterize your analyte [Neutral, Acid, or Base, pKa]
- Select one of the four Oasis sorbents
- Apply the indicated Protocol [1 or 2]
- Determine SPE recoveries by LC analysis

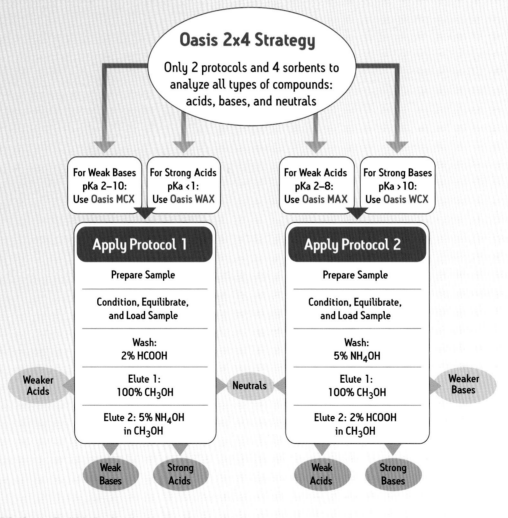

Figure 158: Oasis 2x4 Method

[Oasis Sorbent Technology for SPE]

Oasis Sorbent Selection Tools for Convenient Method Development

Oasis sorbent selection plate and cartridge kits enable rapid development of SPE methods for LC/MS analysis. Having all four Oasis ion-exchange sorbents [MCX, MAX, WAX, and WCX] in a single plate or a cartridge kit is convenient for scouting the best ways to accomplish efficient isolation of unknown analytes, zwitterionic compounds, or mixtures of analytes with different retention/elution properties.

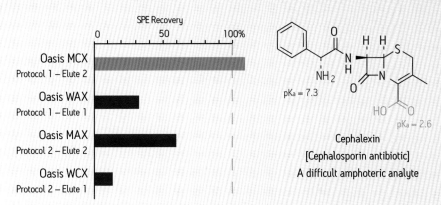

Aliquots of prepared sample were processed using Oasis 2x4 Method protocol designated for each of 4 sorbents. Eluates from Elute 1 and Elute 2 steps were analyzed by LC/MS/MS.
Clearly, Oasis MCX is the sorbent of choice.

Figure 159: Oasis Sorbent Selection 96-Well Plate: Evaluating Oasis 2x4 Method for Cephalexin

Appendix: Applications

[Applications]

SPE Application Examples

In this Appendix, you will find several examples of analytical methods that use SPE. As you can imagine, there are thousands of methods across all sample and molecule types available in peer reviewed literature and on SPE vendors' websites, such as Waters. We have simply selected a few recent applications to highlight the power of this technology.

Determination of PAH in Seafood

Sample Preparation Procedure

Uses DisQuE dispersive sample preparation, Oasis SPE, and Sep-Pak SPE. See full application note for details.

GC Conditions

GC System:	Agilent 6890
Column:	Rxi®-5Sil, 30 m x 0.25 mm, 0.25 μm (df)
Injection Volume:	1.0 μL
Injection Mode:	Splitless (purge time 0.75 min)
Carrier Gas:	Helium
Flow Rate:	0.8 mL/min (constant flow)
Temp. Program:	50 °C initial, hold 1 min, then 10 °C/min to 310 °C, hold 10 min

MS Conditions

MS System:	Quattro micro GC™
Ionization Mode:	EI+

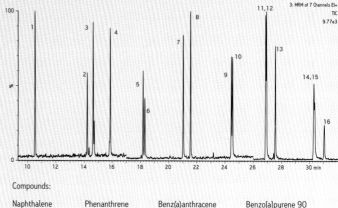

Compounds:

Naphthalene	Phenanthrene	Benz(a)anthracene	Benzo[a]pyrene 90
Acenaphthylene	Anthracene	Chrysene	Indeno(1,2,3-cd)pyrene
Acenaphthene	Fluoranthene	Benzo[b]flouranthene	Dibenz(a,h)anthracene
Fluorene	Pyrene	Benzo[k]flouranthene	Benzo[ghi]pertlene

GS-MS(MS) reconstituted TIC chromatograms of oyster sample spiked at 35 ng/g.

Melamine and Cyanuric Acid in Infant Formula

Sample Preparation Procedure
Uses Oasis SPE. See full application note for details.

LC Conditions

LC System:	ACQUITY UPLC®
Column:	Atlantis® HILIC Silica 2.1 x 150 mm, 3 µm
Part Number:	186002015
Injection Volume:	20 µL
Mobile Phase A:	10 mM ammonium acetate in 50/50 acetonitrile/water
Mobile Phase B:	10 mM ammonium acetate in 95/5 acetonitrile/water

Gradient Table

Time (min)	Flow Rate (mL/min)	%A	%B	Curve
Initial	0.5	0	100	-
2	0.5	0	100	6
3.5	0.5	60	40	6
5	0.5	60	40	6
5.2	0.8	0	100	6
11	0.8	0	100	6
11.1	0.5	0	100	6
14	0.5	0	100	6

MS Conditions

MS System:	ACQUITY® TQD
Ionization Mode:	ESI Positive (melamine and $^{13}C_3^{15}N_3$ melamine)
	ESI Negative (cyanuric acid and $^{13}C_3^{15}N_3$ cyanuric acid)
Capillary Voltage:	3.00 kV

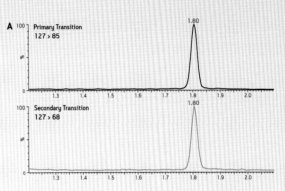
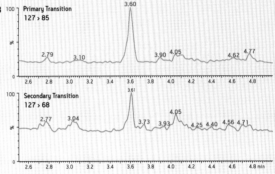

(A) Analyzed by BEH HILIC column, (B) analyzed by Atlantis HILIC silica column.

[Applications]

Multi-Residue Determination of Veterinary Drugs in Milk

Sample Preparation Procedure

Uses Sep-Pak SPE. See full application note for details.

LC Conditions

LC System:	ACQUITY UPLC
Column:	ACQUITY UPLC CSH C_{18}, 2.1 x 100 mm, 1.7 µm
Column Temp:	30 °C
Mobile Phase A:	0.1% formic acid in water
Mobile Phase B:	0.1% formic acid in acetonitrile

Gradient:

Time (min)	Flow (mL/min)	%A	%B	Curve
Initial	0.4	85	15	6
2.5	0.4	60	40	6
3.9	0.4	5	95	6
4.9	0.4	5	95	6
5.0	0.4	85	15	6
7.0	0.4	85	15	6

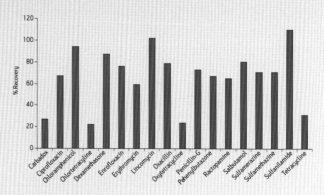

% recovery of 18 veterinary drugs in milk.

MS Conditions

MS System:	ACQUITY TQD

Amyloid β Peptide in Cerebrospinal Fluid

Sample Preparation Procedure

Uses Oasis SPE. See full application note for details.

LC Conditions

LC System:	ACQUITY UPLC
Column:	ACQUITY UPLC BEH C_{18} 300Å, 2.1 x 150 mm, 1.7 µm, Peptide Separation Technology
Column Temp:	50 °C
Flow Rate:	10.0 µL
Mobile Phase A:	0.3% ammonium hydroxide in water
Mobile Phase B:	90/10 acetonitrile/mobile phase A

Gradient:

Time (min)	%A	%B	Curve
0.0	90	10	6
1.0	90	10	6
6.5	55	45	6
6.7	55	45	6
70	90	10	6

MS Conditions

MS System:	Xevo® TQ-S
Capillary Voltage:	2.5 V

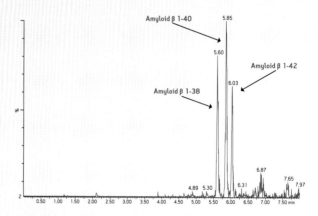

Representative UPLC/MS/MS analysis of amyloid β 1–38, 1–40, and 1–42 peptides extracted from artificial CSF + 5% rat plasma.

[Applications]

Removal of Polyethylene Glycol 400 (PEG 400) from Plasma

Sample Preparation Procedure

Uses Oasis SPE. See full application note for details.

LC Conditions

LC System:	ACQUITY UPLC
Column:	ACQUITY UPLC BEH C_{18}, 2.1 x 50 mm, 1.7 μm
Column Temp:	30 °C
Flow Rate:	0.5 mL/min
Mobile Phase A:	0.1% formic acid
Mobile Phase B:	0.1% formic acid

MS Conditions

MS System:	ACQUITY SQD
Ionization Mode:	ESI+
Capillary Voltage:	2.5 kV

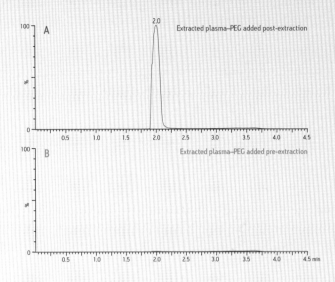

5 mg/mL PEG 400–post-spiked and extracted.

Plasma and Serum Protein Digests

Sample Preparation Procedure

Uses Oasis SPE. See full application note for details.

LC Conditions

LC System:	ACQUITY UPLC
Column:	ACQUITY UPLC BEH C_{18} 300Å, 2.1 x 150 mm, 1.7 μm, Peptide Separation Technology
Flow Rate:	0.3 mL/min
Mobile Phase A:	0.1% formic acid in water
Mobile Phase B:	0.1% formic acid in acetonitrile
Gradient:	100% A to 65% A in 16 min

MS Conditions

MS System:	Xevo TQ-S
Ionization Mode:	ESI-
Capillary Voltage:	3 kV
Cone Voltage:	60 V

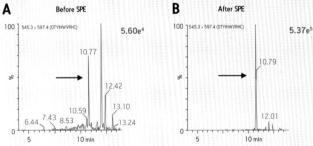

A signature peptide DTYIHWVR detected on a Xevo TQ-S before (A) and after (B) SPE sample cleanup.

[Applications]

Endocrine-Disrupting Compounds in River Water

Sample Preparation Procedure
Uses Oasis SPE. See full application note for details.

LC Conditions

LC System:	Alliance® 2690
Column:	SunFire™ C₁₈, 2.1 x 50 mm, 3.5 µm
Flow Rate:	0.2 mL/min
Mobile Phase A:	Methanol
Mobile Phase B:	Water

Gradient:

Time (min)	Profile % A	% B
0	60	40
10	100	—
18	100	—
20	60	40
23	60	40

MS Conditions

MS System:	Quattro micro triple quadrupole
Ionization Mode:	EI-
Capillary Voltage:	3.2 kV

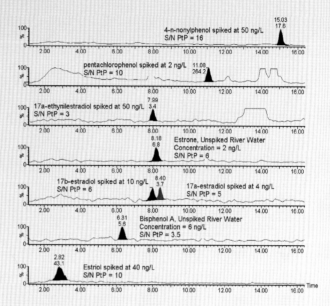

Chromatograms of spiked river water samples.

Organochlorine Pesticides and PCBs in Soil

Sample Preparation Procedure
Uses Oasis SPE. See full application note for details.

LC Conditions See full application note for details.

Compounds:
- U. Unknown
- H. Hydrocarbon (alkane)
- UT. Unknown terpene
- TCP. Trichlorophenol
- BHT. Butylated hydroxytoluene
- UFE. Unknown fatty este
- UFA. Unknown fatty alcohol
- BHB. Butylatedhydroxybenzaldehyde
- BHA. Butylated hydroxyanisole

1. α-HCH (hexachlorocyclohexane)
2. β-HCH
3. γ-HCH (lindane)
4. δ-HCH
5. Heptachlor
6. Aldrin
7. Heptachlor epoxide
8. Endosulfan I
9. DDE
10. Dieldrin
11. Endrin
12. Endosulfan II
13. DDD
14. Endrin aldehyde
15. Endosulfan sulfate
16. DDT
17. Methoxychlor

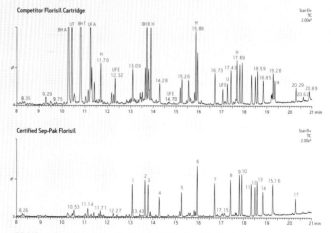

GC-MS analysis of organochlorine pesticides in soil extract.

Appendix: Additional Reference Materials

[Additional Reference Materials]

A comprehensive collection of application notes, journal citations, white papers, books, and other resources are available on waters.com. Resources most relevant to the topic of this book are listed below:

Books

J.C. Arsenault and P.D. McDonald, *Beginners Guide to Liquid Chromatography*, Waters, Milford [2007]; order Part No. **715001531** on waters.com

P.D. McDonald and U. D. Neue, *The Quest for Ultra Performance in Liquid Chromatography: Origins of UPLC Technology*, Waters, Milford [2009]; order Part No. **715002098** on waters.com

E. S. Grumbach and K. J. Fountain, *Comprehensive Guide to HILIC: Hydrophilic Interaction Chromatography*, Waters, Milford [2010]; order Part No. 2531 on waters.com

E. S. Grumbach, J.C. Arsenault and D. R. McCabe, *Beginners Guide to UPLC: Ultra-Performance Liquid Chromatography*, Waters, Milford [2009]; order Part No. **715002099** on waters.com

M. P. Balogh, *The Mass Spectrometry Primer*, Waters, Milford [2009]; order Part No. **715001940** on waters.com

P.D. McDonald and E.S.P. Bouvier, *"A Sample Preparation Primer and Guide to Solid-Phase Extraction Methods Development"*, Waters, Milford [2001]. Search for **WA20300** on waters.com

E. M. Thurman and M. S. Mills, *Solid-Phase Extraction Principles and Practice*, Wiley-Interscience, New York [1998].

U. D. Neue, HPLC Columns, Theory Technology, and Practice, John Wiley & Sons, New York [1997].

White Papers

U.D. Neue, P.D. McDonald, *"Topics in Solid-Phase Extraction. Part 1. Ion Suppression in LC/MS Analysis: A Review. Strategies for its elimination by well-designed, multidimensional solid-phase extraction [SPE] protocols and methods for its quantitative assessment"* [2005]; Search for **720001273EN** on waters.com

Brochures

"Oasis Sample Extraction Products", Waters [2011]; Search for **720001692EN** on waters.com

Catalog

Waters Quality Parts, Chromatography Columns and Supplies Catalog; www.waters.com/catalog